GEOGRAPHY OF MINERALS OF THE OCEANS

GEOGRAPHY OF MINERALS OF THE OCEANS

ATAR SINGH YADAV

Foreword by
Brig. S.S. Grewal SM

CONCEPT PUBLISHING COMPANY, NEW DELHI 110059

ISBN 81-7022-395-4

First Published 1992

Published and Printed by
Ashok Kumar Mittal
Concept Publishing Company
A/15-16, Commercial Block, Mohan Garden
NEW DELHI 110059

Lasertypeset by
Microtech Advance Printing Systems (P.) Ltd.
H-13, Bali Nagar
NEW DELHI 110015

FOREWORD

I HAVE perused through Maj. (Dr.) A.S. Yadav's work on the "Geography of Minerals of the Ocean" and am highly impressed by the findings made by him in the field of Geography and Oceanography. He has justified Sir Halford J. Mackinder's definition of Geography which is the study of hydrosphere.

The importance of oceans and its treasure of minerals cannot be left untapped. Iron and coal are twin pillars of civilisation. The pressure of population, increasing demand of agriculture, railways, heavy and small scale industries and technological advancement of science had compelled the nations to use their natural resources to a maximum limit. Oceans are now not barriers but a thermostat, heat reservoir, provide biotic, abiotic minerals, and cheap transport. A single region of Indian Ocean accounts for 80.7 per cent of extraction of gold, 56.6 per cent of tin, 28.5 per cent of zinc and 77.3 per cent of natural rubber. Other oceans may be having more reserves of minerals.

After introduction the book deals the subject in six chapters followed by a bibliography. The author has described general geography of the ocean in the beginning chapter. The next chapters describe biotic and abiotic minerals of the world. A brief history of the oceanography in develop and under-developed countries, exploitations programme of India at Antarctica and trade routes of minerals through Suez and Panama canals are described in subsequent chapters. In conclusion, appendices and bibliography simplify the growing science of oceanography.

Indeed the book has a number of innovative features. It speaks of the ingenuity on the part of the author to interpret material with the help of facts and figures. The language is simple, clear and easy to understand by a reader.

I am sure that this study will go a long way in exploring and exploiting enormous potential reserves of the oceans with greater understanding and cooperation.

Brigadier S.S. GREWAL SM
Commander of an Infantry Bde

ACKNOWLEDGEMENTS

I WISH to express my deep sense of gratitude to Dr. Madhu Sudan Singh (Retd.) Principal and Head of Deptt. of Geography Raja Balwant Singh College, Agra, Brig. S.S. Grewal SM, Lt. Cdr. D.S. Dalal, Prof. Dr. I.N.G. V.S. Raju, Dr. M. Ravindran, Dr. S.P. Subramanian, Dr. K. Genesh Babu of Ocean Engineering Centre, Indian Institute of Technology, Madras, Dr. A.K. Malhotra formerly member of Oil and Natural Gas Commission, Bombay, Prof. G.L. Nanda for sending me their original articles about geographical structure of oceans, deep sea mining, ocean resources and offshore activities related to oil and gas in India. Without their assistance I could not have completed this work. It was the dissertation of Brig. S.S. Grewal SM about 'Exploitation of Resources of Ocean' which inspired me to compile this book.

I would also like to express my sincere gratitude and regards to officers of the Department of Ocean Development, Government of India, New Delhi, for sending me valuable material, articles, pamphlets and reports about ocean science and technology, India and Antarctic Scenario, marine science promotion of research and development activities.

In the same way, I should express my deep gratitude to the Head of the Library, Panjab University, Chandigarh, Librarian Jawahar Lal Nehru University, Delhi, Librarian Institute for Defence Studies and Analysis, Sapru House, New Delhi for providing the required material about this topic. Mr. S.Z. Qasim, Secretary of the Government of India, Department of Ocean Development, South Block, New Delhi-110011, and Department of Fisheries Government of India, deserve my special thanks for allowing me some data from published book article. "A Technological Forecast of Ocean Research and Development in India".

I wish to thank Col. Moti Singh, Col. Man Mohan Singh, Maj. K.K. Chopra and Maj. Arvind Govil and many other friends of Armed forces of this formation who have rendered great help in assisting me in compilation and preparation of maps for this book.

In the end I am highly thankful to a large number of persons, officials and non officials who not only provided good suggestions but also readily gave information from their empirical knowledge.

Finally I express my thanks to Sub Shree Ram, Personal Assistant to Cdr. for typing this work. Services of Sub Badesra will also be remembered by me for his valuable assistance.

MAJOR (DR.) A.S. YADAV

CONTENTS

LIST OF FIGURES

LIST OF TABLES

INTRODUCTION

"The ocean form the most prominent frontiers features of the planet."

Arthur A. Halls Essay on
"Boundaries in International Relations"

Historical Background

THE OCEANS form the most significant natural frontier of the nations. The earth as a planet has 71 per cent of water cover and 29 per cent of land. The resources biotic or abiotic of land are being exploited at a rate where their reserves will not last for a long period. There is a increasing pressure of population over these resources. The sea, in near future would become increasingly significant source of minerals and energy. The technological advancement has reduced the distances, depth of high seas. Sailors, scientist, marine engineers and naval experts have started exploration and exploitation of the resources available in adjacent continental shelves, continental slopes and ocean beds.

The early voyages of discovery such as those of Columbus, Magellan and Cook extended man's geographic knowledge of the ocean. The first serious attempt to explore the ocean body and ocean floor was made by the British Challenger (1872-1876). This global voyage revolutionised the attempts to explore ocean deposits and oozes. This marked the beginning of the modern oceanography. Echo sounding developed in First World War period (AD 1914 to 1917) which helped in the measurement of depth and led to the making of new maps of the continental shelves and slopes. Oceanographic institutions began to develop in the 1920s and 1930s.

The real progress in the field of oceanography and marine sciences began to develop after World War II. The pioneer countries in the field of oceanography are USA, USSR, followed by Great Britain, France, Japan, West Germany and others. A great advancement in ocean exploration came with the development of deep diving vehicles equipped with portholes. By mid of 1970 about 50 such submarines were reported operational in the world. The modern scientific advancement developed deep diving and drilling capabilities which enhanced geological knowledge of rocks history and availability of minerals. The research ship Glammer Challenger has penetrated ocean sediments for thousands feet. Scientists from many countries participated in drilling voyages.[1]

Primarily the sea is helpful to mankind in numerous ways. It acts as a great thermostat and heat reservoir, levelling out the temperature extremes which would prevail over the earth without its moderating influences. It is earth's water reservoir and also provides the least expensive form of transportation. From the beginning, man has also extracted riches from the sea for his use.[2] India has to revise its planning for harnessing the off-shore riches since she has a long coastline and the United Nations conferences on "Law of the Sea" has extended the economic jurisdiction of the coastal states to an area ranging from 200 to 300 mm from the low water line. This has given India exclusive rights to operate in area up to 1,50,000 sq km in high seas which would be equivalent 60 per cent of the total country's land mass.

Initial survey has indicated that Indian coastline and continental shelf contains polymetallic nodules, the source commercially imported metals such as manganese, nickel, cobalt and to a lesser extent metals like molybdium, vanadium, zinc, lead and cadmium scattered on the ocean floor at depths ranging from 3500 to 6000 metres. India has been granted exclusive sea bed mining rights and earmarked successfully on the production of oil from the off-shore regions of Arabian Sea like Bombay High and Bay of Bengal, Cauvery Basin. India could hope to mine three million tonnes of these nodules a year within next decade.[3] These polymetallic nodules contain 15,000 tonnes of nickel, 1200 tonnes of copper, 60,000 tonnes manganese, 1500 tonnes of cobalt and undefined quality of gold. The Indian Ocean in total has an estimated 150 million tonnes of the polymetallic nodules.[4]

Statement of the Problem

An attempt is made in this investigation to describe the "Geography of the Minerals of the Oceans" with reference to developed technology and exploitation programmes of India in future".

Justification for the Study

The investigator has selected this topic because very little work has been done by marine science engineers, scientists, geologist and oceanographers of under-developed and developed countries. There are manifold fields of ocean, i.e. physical and chemical oceanography, marine biology, marine geology, marine geophysics, ocean engineering, marine ecology, meteorology and marine instrumentation which need scientific exploration and exploitation of minerals in near future.

A new regime of the oceans has come into being in December 1982, when at Jamaica, more than 100 countries affixed their signatures on to a new convention to govern the wealth and uses of the seas. These new laws hold an immense promise of development particularly for the developing nations in as much as their rights and sharing of ocean wealth are protected. In such an important situation, India has a new role to play and a responsibility to meet for both exploration and exploitation of the vast ocean located at its door step. As a first measure in this direction, a new Department of Ocean Development was created on 24 July, 1981, directly under the late Prime Minister Mrs. Indira Gandhi. Since then, this department has played a key-role in effectively projecting India's role in all International meetings, in sea bed research, survey of polymetallic nodules and in the organisation of successful expeditions to Antarctica. Being a new department of ocean development its activities should be described and studied in details as most of the individuals don't know much about ocean resources of living organisms and non-living minerals.

There is a growing pressure on the exploitation of resources of land due to increase in population and enormous demand of industrialisation in India as well other countries. The demographers agree that population of India will increase to 1000 million towards the end of this century i.e. 2000 years A.D. Others estimated that by the year 2000 A.D. the population of India will be around 998 million, of this 508 million will probably be males and 490 million females. This population exerts enormous pressure over the land resources of India. The total land area of India is 327×108 km^2 which is only 0.64 per cent of the earth's surface. This would also mean that about one sixth of the world population would become congregated in a 150th part of the world's surface area and this population will have to be sustained by the agricultural land of about 1.65 × 10^5 km^2. Another source of resources will be the adjacent Indian Ocean with Arabian Sea and Bay of Bengal. India has a coastline of about 6000 km. Assuming that 25 per cent of the present population of 680 million live in the coastal areas (within a few hundred miles from the sea shore), it can be estimated that about 170 million people are directly dependent on the sea for their living or livelihood. It becomes indispensable to study these resources of the Indian Ocean.

The Gulf countries have increased the prices of petroleum and crude oil. There has been increase in consumption of use of petroleum products in mechanised transport and various industries. The land reserves of crude oil are concentrated in North East Region of India and are insufficient to meet the national demand. Indian Ocean oil explorations of Bombay High, Cauvery, Godavary and Krishna Basins hold greater prospects of oil reserves for India.

In order to understand applicability of advanced technology used for extracting polymetallic nodules from high and deep oceans this study has become essential for defence, commercial agencies, scientist and marine engineers. All have to accept the challenges of oceans with the developed sophisticated scientific technology of hydraulic lifting, air lifting and continuous bucket system. Large scale mining of ocean bed in a surging, moving water mass calls for high skills and most advanced technology. India is using 'Sagar Sampada' and 'Sagar Kanya' oceanographic research vessels for the enhancement of this knowledge and exploitations.

Besides, the above India shares the status of a pioneer investor along with USSR, France, Japan and four multinational consortia. It has its programmes of Antarctica Expeditions. The first Indian expedition landed at Antarctica on 09 January, 1982. Other expeditions established a well equipped permanent station named '*Dakshin Gangotri*' at latitude 70° 05′ and longitude 12° 00′ E. India had to invent in more underwater technology to exploit treasures of oceans as well as Antarctica Ocean.

Sources

The main sources of this work are articles, pamphlets and books published. Published pamphlets of Department of Ocean Development, Government of India, New Delhi, articles obtained from Ocean Engineering Centre, Indian Institute of Technology, Madras, articles published in Strategic Digest, Institute for Defence Studies and Analysis have helped the author in collection of the data. Dissertations submitted to University of Madras about exploration of ocean depth also broadened the scope of the subject-matter. Various books pertaining to area of ocean, marine researches and oceanography, marine biology and ocean expeditions assisted the investigator in compiling.

Methodology Used

As regards the methodology and technique of dissertation a very comprehensive study of the literature connected with past historical explorations of oceanography, recent developments in technological researches, Government efforts to exploit the resources of ocean has been done by the investigator. It is basically a normative research following historical method. The published material records, reports, letters, articles, books and pamphlets issued by Department of Ocean Development, Government of India, New Delhi. Books available at Punjab University, Chandigarh, Jawaharlal Nehru University, Delhi, Delhi University, Delhi, Sapru House, New Delhi, were consulted and data obtained has been arranged systematically. Graphs, maps and tables were procured and arranged in order of chapters.

NOTES

1. *Merit Student Encyclopedia* p. 14 Vol 14, Macmillan Educational Corporation, New York, 1979.
2. Article of Dr. M. Ravindran, "Deep Sea Mining" p. 1, Ocean Engineering Centre, I.I.T., Madras.
3. Editorial in the *Tribune*, June 22, 1982.
4. Lt. Cdr. D.S. Dalal, N.M., "Exploration of Ocean Depth — Recent Development in use of Under Water Robots and Suitable Designs for Indian Conditions" Dissertation submitted to University of Madras, January, 1988.

1

GEOGRAPHY OF THE OCEANS, ECONOMIC LANES OF NATION WITH SPECIAL REFERENCE TO INDIA

"Hast thou entered into the spring of the sea or hast thou walked in the search of the depth."

—Job 38

"Water is everywhere the sovereign ruler of human activity."

—Jean Brunhes

THE SOURCE of water is the oceans. No other planet in the solar system has any surface water at all as far as we know, whereas the Earth's great oceans and seas cover seven-tenths of the globe.[1] It is believed that most of the water in the oceans came from the earth's interior, brought to the surface by volcanic activity. Whatever its origins, water covers about 71 per cent of the earth.[2] The dominance of water on the surface of our planet is revealed by comparisons of land elevation and oceanic depths. The average height of land is 850 metres above sea level. The average ocean depth is 3730 metres. The highest terrestrial mountain, Everest rises over 8800 metres above the sea surface, while the deepest area of the ocean floor the Marinas Trench in the Western Pacific, drops to 11000 metres below sea level. Did you know that highest mountain on the earth lies not in the Himalaya but in the Hawaien Island Chain.[3] Mauma Kea ascends 9763 metres from base to peak, but over half of its elevation lies under water. And consider this; if erosion to wear down all the surface irregularities of the earth and uniformly redeposit the resultant sediment, a global ocean nearly 2-$^1/_2$ km deep would completely cover the planet. This can be easily understood by a graph described by Henry S. Parkar. Oceans are not composed of uniform masses of water but of a series of layers differing in temperature and salt content with each level supporting its own form of marine life.[4] Before the description of ocean resources one should know about oceans and seas of the world.

THE PACIFIC OCEAN

The Pacific, in area some 165,384,000 sq km, is the largest of the Oceans of the world. It is more than twice the size of the Atlantic, the next largest ocean, and occupies nearly half the Earth's surface. In the North it is almost land locked, its only outlet to the Arctic Ocean being through the Bering Strait. It stretches 15,216 km from this strait, to the Antarctic and at its broadest is 16,885 km across. It has an average depth of about 4300 metres but in its deepest regions, which are off the Philippine Islands a depth of 10,912 metres has been recorded. The Islands of the Pacific fall into three main groups, Micronesia, Malanesia and Polynesia. Nearly all these islands are either volcanic or have capping of coral reefs over a submarine volcanic peak.[5] It is almost circular in shape.

THE ATLANTIC OCEAN

The world's second largest ocean, has the same name as the Atlas Mountains, derived from Greek God Atlas.[6] It has an area of 32 million square miles which is 10 times the size of the United States of America. It separates America (The New World) from Europe and Africa (The Old World). Shaped like an hour-glass, the Atlantic is narrowest (less than 2000 miles wide) between the western bulge of Africa and the north eastern coast of Brazil. At the latitude of the United States of America and Western Europe, it is about 3000 miles wide, a distance which the fastest ocean liners can cover in less than four days. The greatest depth recorded is the Milwaukee Deep (30250 feet), off the island of Puerto Rico in the Caribbean Sea. From North to South in the centre of Atlantic Ocean, is a submarine ridge on which several islands are sited : Iceland, the Azores, Ascension Island, Tristan da Gunha, Gough Island and Bouvet Island.[7]

INDIAN OCEAN

The Indian Ocean, the largest in the world after the Pacific and Atlantic is the part of the ocean world that lies South of Asia between Africa and Australia with Antarctica in South. It joins the Atlantic Ocean South of the Southern tip of Africa; their border runs from Cape

Agulhas (at the southern tip of Africa) to the South along 20°E meridian up to the shores of Antarctica. In the East and South East, its waters mingle with the Pacific. In the South-East, the border runs from South East Cape in the Island of Tasmania and Australia, it includes the Bass Strait, however, some researchers consider Bass Strait as part of the Pacific. In the North-East, it runs along the Torres Strait between Australia and New Guinea, then from the Island of Adi, off the Western Guinea, along the Southern shores of the lesser Sunda Island and the Island of Java, thence across the Sunda Strait to the shores of Sumatra and then to Singapore.[8] The Northern most limit of the ocean reaches up to 30°N and to include Red Sea and the Persian Gulf. Looking like an inverted "W", the Indian ocean has two branches, the Arabian Sea and Bay of Bengal. The unique feature is that, unlike the Pacific and the Atlantic, it is not open ended, that is it does not extend from Pole to Pole. It is the only embayed ocean, most part of its area is walled off on three sides by land with the southern side of Asia forming a roof on it. It does not extend in the North to cold climate regions, it remains navigable throughout the year.

Its total area, including the Persian Gulf and Red Sea is estimated to be about 28,400,000 square miles (73,600,000 sq km). Its mean width in the South between the Southern point of Africa and Australia is more than 6200 miles (10090 km) and it narrows steadily northwards the width between Africa and the West Coast of Australia and Indonesia being 6400 km to the South. Its average depth is 12,706 feet (3,890 mtrs) and it contains an average of 70,086,000 cubic miles (292,131,030 cubic km) of water.[9] In comparison with the other oceans, the Indian Ocean has the fewest seas. Indian Ocean forms a large triangle with the Indian Peninsula protruding the upper apex.[10]

THE ANTARCTIC OCEAN

The Antarctic Ocean comprises seas and lands around the South Pole which lie within the Antarctic circle 66°33′ S. It has an area of about 11.7 million sq km (4.5 million sq miles). The continent is uniquely isolated and is covered by an ice cap thousands of metres thick. Much of the rock surface beneath the ice is below sea level.

THE ARCTIC OCEAN

It has an area of 1,40,89,600 sq km and depth ranging to 5441.

Oceans of the world cover 90 per cent area in Southern Hemisphere and 10 per cent area in Northern Hemisphere. Details of area volume and mean depth of all the oceans are described in Appendix 'A' to the dissertation.

The ocean floor is divided into four[11] important physical divisions.

Continental Margin

It is the part of the ocean adjacent to the continents located between emerged land and deep sea. It includes the continental shelf, the continental slope and the continental rise.

Continental Shelf (Littoral)

Continental shelf is the area of the ocean floor between mean low-water line and the location of marked change in the inclination, of the floor at the upper edge of the continental slope. The change in inclination, from average 0-1 degree to 3-2 degrees occurs at depths usually between 130 and 200 metres, but exceptional cases as shallow as 50 metres. The width of the shelf ranges from about 1 km to 1200 km.

Continental Slope

It is usually 15 to 30 km wide, extends from outer edge of the continental shelf to the continental rise. The inclination of the slope varies widely from as little as 3° to over 45°, slopes of 25° are common. Depth is 2000 fathoms.

Continental Rise

Inside the ocean it is a broad, uniform and smooth surfaced wedge of sediments up to 10 km thick, is 100 to 1000 km wide and it slopes gently oceanwards from the base of the continental slope, usually in 2000 to 5000 metres of water.

The continental margin's size is highly significant, with an estimated area of 74.5 million square kilometres, which is about 53 per cent of the total emerged lands.

TABLE 1.1: Relative Size of Emerged Lands Continental Margins and Deep Sea Beds

Size of Ocean Floor	Area	% of total area of
1. Emerged lands	141 Million km^2	27.0
2. Continental shelves and slopes	55 Million km^2	10.8
3. Continental rise	19 Million km^2	3.8
4. Sub-total continental margins	74 Million km^2	14.6
5. Deep sea bed	285 Million km^2	57.8
Total World	510 Million km^2	100

The proposed exclusive economic zone of 200 nautical mile limit has an area of 120 million square kilometres and represents 35 per cent of the total ocean and about 86 per cent of the emerged lands. These areas are shared by 130 coastal countries, out of which 75 per cent are developing countries. The following details show a step-wise form of the topography:

(a) Summit – All land above 100 m
(b) Continental Plateau – Land below 1000 m and continental shelf to 200 m
(c) Continental slope – From the edge of the shelf to the mean crust level of 2440 m
(d) Deep sea bottom – From 2440 m to 5750 m
(e) Deep sea depression/ trenches – Below 5750 m

Economic Lanes of a Nation with Special Reference to India

With the advent of scientific technology oceans have become a source of food, energy, raw material, commercial and naval trade. Admiral Gorshkov's has aptly remarked, "The constantly growing maritime might of our country ensures our ability to enlarge our exploitation of the colossal natural resources of the world ocean".[12] Over the last 30 years uses of the ocean have increased in a dramatic fashion with the growing, realisation of limits to land – based resources and the expansion of world trade. Between 1950 and 1974 annual world fish harvests rose from 16 to 67 million tonnes, and world shipping from 76 to 306 million gross registered tonnage of all ocean uses, resources of the sea bed have shown the largest expansion from virtual non-existence to the present high level which will continue to increase in the foreseable future. This has been made possible by new technology.[13]

Before analysing various agreements, treaties and acts passed and accepted by coastal nations of the world it is pertinent to know the international frontiers and zones of various nations of the world. It was the Artillery range of the bomb which determined coastal frontiers of economic lanes and international jurisdiction over the sea. At the time of independence India accepted three miles low tide zone as international frontier along with Pakistan. When Pakistan enhanced its frontiers to six nautical miles India had to revise and extended her territorial boundaries to six nautical miles. With the improvement of Artillery range Pakistan and India had extended their coastal boundary to 12 nautical miles. The coastal boundary of other countries ranges from 10 nautical miles to 200 nautical miles. A small country like Chile has declared its coastal boundary up to 200 nautical miles. The U.S.A. and Canada claim their right of exploitation of minerals in the ocean depths, deep seas and in the entire bed of the oceans adjacent to their countries (see Appendix B for various claims of territorial waters). Exploitation of resources of sea bed has been confined so far to those in relatively shallow waters, mostly less than 200 m on the continental shelf in close proximity to land.[14]

Law of the Sea and Economic Lanes

The increase in importance of resources of the sea has resulted in conflicts of interest, which has even reached the stage of armed clashes. So in pursuit of world harmony considerable pressure has been brought to bear for the revision of traditional grotion freedom of the seas. Historically the movement started in 1945 with the unilateral declaration of president Truman that the U.S.A. had exclusive right to explore the resources of continental shelf beyond national territorial seas, followed by further claims by South American countries of jurisdiction out to 200 miles from the coasts. The need to reconcile these claims with the law of the sea led to the first United Nation Law of the Sea Conference (UNCLOSI) in 1958. It gave states sovereign rights to explore and exploit natural resources of their adjacent continental shelf to the water depth of 200 m.

In 1960, a second UNCLOS was convened to resolve residual difficulties, but it made no progress. Further impetus was given to

Pardo's declaration in 1967 followed by a multitude of unilateral claims over all aspect of sea use. Finally in 1973 third UN conference (UNCLOS III) was launched and has been making slow progress.

(a) A general concensus reached on agreement on the 12 mile width of territorial sea giving exclusive jurisdiction over the area as though an extension of the land mass of the state.
(b) The second main advance has been the overall agreement on the principle of a 200 mile, Exclusive Economic Zone (EEZ) in which coastal states would have sovereign right and jurisdiction over natural resources. The 200 mile EEZ comprise 35 per cent of the world ocean area.[15]

Further technical advances in exploitation of ocean resource are likely to lead an increase in conflicting uses. The Indian ocean has an exceptional importance for economic, political and strategical use. The area of the Indian Ocean is rich in several important mineral,[16] raw material and marine food and has been regarded world's major store-house of natural resources. The region particularly accounts for 80.7 per cent of world extraction of gold, 56.6 per cent of tin, 28.5 per cent of manganese, 25.2 per cent of nickle, 18.5 per cent of bauxite, 12.5 per cent of zinc and 77.3 per cent of natural rubber. The importance of Indian Ocean tremendously enhances in view of the large oil bearing regions on its coast. Important commercial sea lanes lie across Indian Ocean which function as main connection link between the Atlantic and Pacific Oceans; the routes passing through it connect Europe and East Africa with South and South East Asia, far East and Australia. It has become a very important area of navigation and marine trade. One-fourth of entire cargo of marine trade and two-thirds of oil are loaded and unloaded in the ports of this ocean. About 60,000 vessels pass through it every year.

Ocean have greater significance for controlling climate, marine fisheries, navigation and from the point of abiotic resources. Pacific, Atlantic, Indian, Arctic and Antarctica are the vast oceans besides the seas in the world. With the increase of technological advancement their significance for resources increased. There are many claims to the territorial oceanic boundary by coastal nations. Life and development in India cannot be thought without the influence of Indian Ocean which is supposed to be the future store house of minerals, food and navigation.

NOTES

1. The Reader's Digest Great World Atlas p. 118. The Reader's Digest Association, London.
2. Readers Digest Library of Modern Knowledge No. 1. The World of Nature p. 100.
3. Henry S. Parkar "Exploring the Ocean" pp. 38-39 Prentice Hall Inc Englewood Cliffs, new Jersy — 07632.
4. The Readers Digest Great World Atlas p. 118.
5. *Ibid.* pp. 98, 99.
6. Hamlyn, "Encyclopaedia of the World in Colour" p. 41. London 1973.
7. *Ibid.* p. 41.
8. Lt. Col. (Retd.) Dr. Bhupinder Singh, 'Indian Ocean and Regional Security' p. 1 — 1983 B.C. Publishers 47, Ajit Nagar, Opposite Fatima Convent Patiala-147001 Punjab (India).
9. *Ibid.* pp. 2, 3.
10. Editor, William P. Hlsey, "Merit students Encyclopedia", Vol 14 p. 7. Macmillan Educational Corporation. New York.
11. Prof. Dr. V.S. Raju "Introductory Lecture on Ocean Resources and Ocean Engineering Aspects". PP. 1, 2. Ocean Engineering Centre, IIT Madras 36.
12. S.G. Garshkov, "The Sea Power of the State".
13. Article by Lieutenant Commander M.P. Graton, Royal Navy, "Protecting New Frontiers on the Ocean Floor" p. 32, Published in RUSI Journal of the Royal United Services Institute for Defence Studies Italy — U.S. — Canada, March 1979.
14. *Ibid.* p. 32.
15. *Ibid.* p. 33.
16. Lt. Col. (Retd.) Dr. Bhupinder Singh, "Indian Ocean and Regional Security" p. 3.

2

GEOGRAPHY OF BIOTIC RESOURCES AND MINERALS OF OCEANS

"And God said, let the nature bring forth abundantly the moving creature that hath life".

Genesis

IMPORTANCE IN GENERAL

THE SEA, in near future, would become increasingly important source of energy and minerals as the land deposits dwindle. Until recently the submerged wealth were unavailable because the know-how and technology suitable for exploration of mineral did not exist. This situation has changed in past two decades. Man has begun with offshore and deep sea drilling to tap the riches of ocean floor. The bottom has always been rich in deposits of minerals. Oceans have both biotic as well as abiotic resources.

THE RESOURCES

A resource has been defined as any means of attaining given ends. A resource satisfies human wants.[1] Water, air, sunshine, land, oil, forest, wildlife, fishes, minerals and power are resources all useful to man. Here one has to understand element too. It is a single substance found in atom having it's peculiar nature. Aggregate of elements is known as mineral. A mineral is definite chemical composition, comprising one or more elements and can be identified by its physical and chemical properties. At the same time aggregate of many minerals and living organism and sources of energy of the earth and sea became resources. Recently the concept of natural resources has been broadened to include the total natural environment that is, the entire surface layer of the earth as well as ocean. From this all living and non-living elements of the atmosphere, the oceans, the deserts, the polar, tropical and temperate region and the earth too have all become valuable resources.

The resources are classified into the following types:

(a) Biotic and Abiotic.
(b) Exhaustible and inexhaustible.
(c) Potential and developed resources.
(d) Agricultural and pastoral.
(e) Raw material and energy resources.
(f) Mineral and industrial resources.

In this dissertation the investigator will use the Biotic and Abiotic classification of resources.

Biotic Resources

The most clearly recognisable natural resources are those consisting of living things i.e., forest and their products, agricultural crops, animal fodder, wild and domestic animals, birds, reptiles and fishes. They can continue to reproduce and regenerate their population as long as environmental conditions remain favourable. The biotic resources of oceans include fisheries, mariculture, seaweeds, mangroves and coralreefs.

Abiotic Resources

The abiotic resources consist of non living things. In general they are considered mostly non-renewable. Minerals and fossil fuel such as coal, petroleum and natural gas are abiotic resources. Iron, aluminium, gold, silver, platinum, manganese ore, cobalt and nickle are abiotic to a greater extent.[2] Lime, sand and gravel, barite (barium sulphate), iron sands, tinoxide, diamond, phosphorite platinum ilmenite, zircon, rutile, titanite, cassiterite, monozite, chromite potash, sulphur, salt, salt bring and fresh water are also included in this category.

Let us investigate the exploitation of biotic resources of the oceans. The sea, where life began, contains representatives of almost all the main groups of animals including insects of which close to one million species are known, the species of marine animals numbering about 16,000 would be twice as many as those on land.[3] In the see life depends largely on plants which are equally productive. The pastures of the sea and the basis of its life cycle are countless myrias of free floating, microscopic plants known as phytoplankton. These are the food of minute animal called zooplankton. Life in the sea depends largely on the productivity of microscopic chlorophyll bearing plants, such as diatums, various flagellas and blue green algae. These plants form part of plankton and flourish where sunlight is strong enough for photosyntheis which in the open oceans is between the surface and a depth of about 300 feet. Beneath the sunlight 300 feet layer, midwater animal life becomes more and more sparse as the depth increases. Many of the animals that live between zone of 500 to 3000 feet deep migrate up to the surface water each day in search of food. Luminous crustaceans squid and fish are among such migrants.[4]

Capture of Fishery

Fisheries mean the capture and processing of sea, coastal and inland aquatic animals as an occupation for profit. There are about 30,000 species of fish, many of which live only in fresh waters. Sea water is important both for local demand and for wider demand outside. The zone of maximum important to mankind to day for the exploitation of natural living resources is the upper most layer of 100 meters depth of ocean. It forms 1 - 8 per cent of the world oceans. From this zone, more than 50 per cent of the worlds fish catch is obtained. The regions occupying this zone are either fairly close to the coast or in very fertile areas of the coastal or offshore upwelling regions called 'oasis' in the sea. Since such area constitute only 25 per cent of the ocean area, it can be assumed that 75 per cent of the ocean area may be termed as oligotrophic with moderate to low production rates.[5] Fish are a vital source of food specially in countries like Japan, Norway, Iceland and Newfoundland where land is mountainous. The productivity of ocean fishes is affected by many factors e.g. the depth of the water, the ocean current, the temperature and salinity determine the amount of plankton (rich food) present.[6]

Principal Fishing Ground of World

Although fish may be caught in the sea and in fresh water in any part of the globe. Figuré 2.1 shows that the worlds major commercial fishing grounds are located in the cool waters of the northern hemisphere in comparatively higher latitudes. Commercial fishing is little developed in the tropics or in the southern hemisphere. There are three major fishing grounds (Fig. 2.1) in the world which are all located in the temperate latitude of the northern hemisphere. The major fishing grounds are:

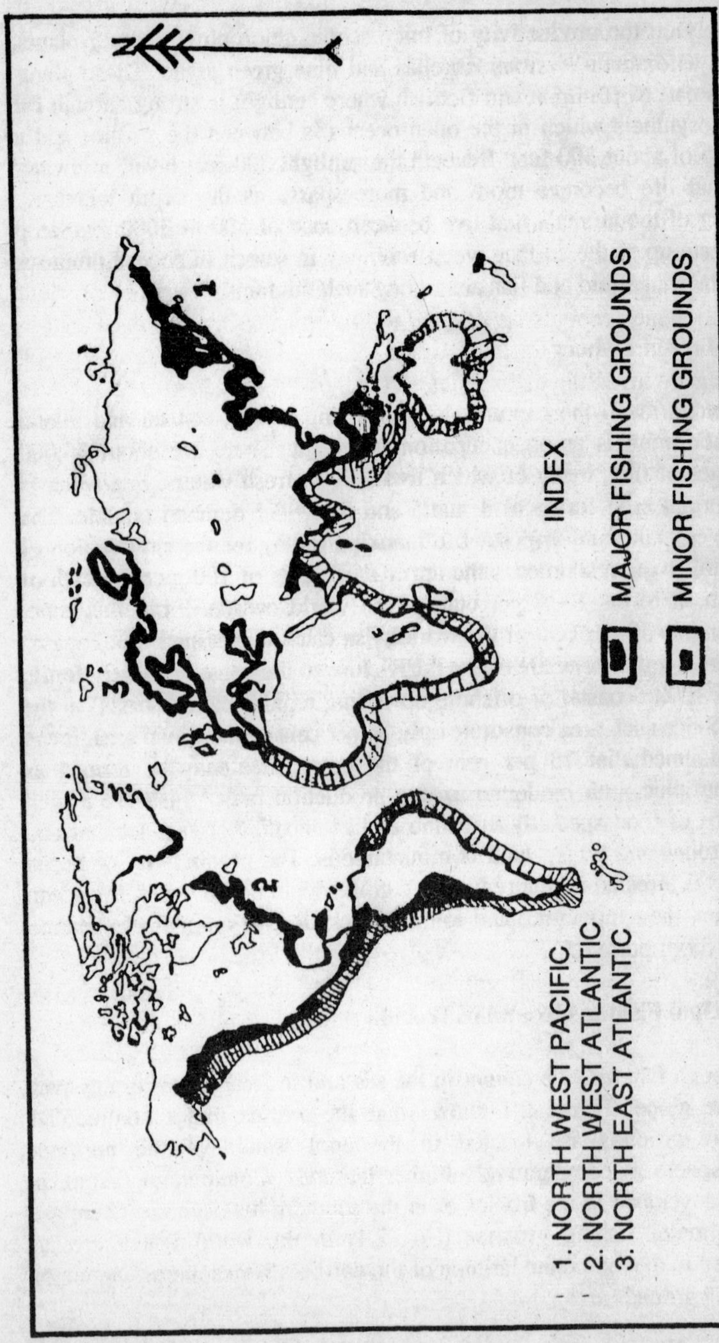

FIG. 2.1: Principal Fishing Grounds of World

(a) The North-West Pacific
(b) The North-West Atlantic
(c) The North-East Atlantic

The North-West Pacific Fishing Grounds

The North-West Pacific from the Berring Sea to the East China Sea is the worlds greatest fishing region (Fig. 2.1). Computed in total tonnage of fish caught the annual yield is well over 10 million tonnes of which Japan alone accounts for about 7 million tonnes. The rest is shared by China, Korea and Eastern Russia. There are probably more people engaged in fishing here than in any other part of the world. Commercial fishing is best developed in Japan where the industry is highly organised and the fishing method are modern.[7] The North West Pacific fishing ground has the geographical advantage of location in the Northern Pacific area, the confluence of Kuro Shio warm water with the cold water of the Kuril current, the scientific management of fisheries and the great demand of fish in the domestic market have made Japan one of the leading countries of the world in production and catch of fisheries. From the Japan Sea the Japanese capture heavy hauls of sardines, herring, mackerel, sharks solomon, yellow tail, cuttle fish, shell fish and crustance only of which crabs and lobstes are the most important. Deep sea fishing for cod, tuna, bonitos, turtles, mackereland halibut has developed more recently. Japan consumes more fish than any other country in the world.

The North-West Atlantic Fishing Grounds

This fishing ground stretches between New Foundland and New England (U.S.A.). The New Foundland waters are a meeting place for the warm current and the cold current namely Gulf stream and Labrador respectively, which is favourable to the development of swarms of plankton (fish food). Beside, the presence of 'banks' or submarine plateau, grooved by ice erosion and littered with moraines, locate one of the most productive fishing grounds in the Atlantic. The New Foundland fishing grounds have been visited by the fishing fleets of Britain, France, Spain and Portugal and these have now been joined by those of Japan and U.S.S.R. Besides cod, herring and mackerel are caught in the surface waters, halibut, haddlck, hake and flonder in the deep waters. In New Foundland fishing provides employment for ninety per cent of the population. Fish in all its various forms fresh, crude, preserved or canned, is sent in large quantities to mainland America and exported to Latin America, Southern Europe and North Africa.

The North-East Atlantic Fishing Grounds

The North-East Atlantic fishing region extends from Iceland to the Mediterranean shores (Fig. 2.1). The North Sea is the largest fishing ground in the world. It is very shallow and abounds with fishing banks (Dogger and the Great fisher Bank). The countries surrounding the sea i.e., Great Britain, France, Northernlands, Belgium, Denmark, Norway, Sweden, F.R.G., G.D.R., Ireland and Iceland are densely populated providing a good market for catches.[8] The principal fish caught are herring, cod and mackerel. In the colder northern waters, haddock, turbot and halibut are caught and hake, skate, plaice and sole are also important. Further south anchovies, pilchards and sardines are the main species and some tuna are also found. The North East Atlantic region with an annual catch of about 9 million tonnes is the largest fish exporting region in the world.

The North-East Pacific adjacent to shores of North America from Alaska to California is also a rich fishing region. In the area, Salmon is the most important fish caught. It is mostly exported as canned Salmon. Halibut, herring, tuna, sardine, crabs, shrimps and oysters are also caught for the North American markets. As regard tropical waters though exploited on a small scale for countries have less fishing potentials because fish of commercial species are fewer. Tropical seas are richer in species than the colder seas, but they are poor in the quantity of fish and they, therefore, do not have many important centres of high sea fishing. Moreover, there are a few wide continental shelves except South East Asia and Coastlines are often straight and unsheltered. In the tropical latitudes fish species include redsnapper, mullet, sea trout, flying fish, spanish mackerel, drum and flander.

Fishing in India

Fish is an important stem in the diet of people living in the coastal states of India. India has a long coastline, yet only about 8 lakhs tonnes of marine fish are caught annually. See fisheries are confined to the coastal waters from the shore in Gujarat, Malabar coast, Gulf of

Manner, Tamil Nadu coast and the waters over narrow belts of continental shelves and slopes. In 1947, fish production consist largely of capture fisheries and for those the intensively exploited areas, are found in the narrow coastal belt. In 1947 fish production in India was about 0.4 million tonnes. In 1979-80 it had increased to 1.4 million tonnes. In 1970, it had passed the million mark and thereafter its increase has been somewhat unsteady. The estimates of potential fish yield from the Indian Ocean vary from the about 7 million to 17 million tonnes of this potential, India's contribution is expected to be of the order of 5 - 9 million tonnes. The fish production from the Indian Exclusive Economic Zone (EEZ) is expected to be about 3 million tonnes. The projected fish requirement of India's population by the year 2000 are estimated to be 11.4 million tonnes of this 60 - 75 per cent is expected to come from the sea and the rest from inland waters. The varieties caught along the Indian shores are herring, mackerel, prawns, catfish, mullets, pomfrets, Indian salmon, jawfish, flounders, halibut, shark, hakes and haddock etc. (See further details in appendix-C, D and E.)

Mariculture

It is being practised on a small scale in the enclosed back water and esturine areas of Kerala, Karnataka and West Bengal. Based on traditional method of trapping juveniles of prawns and fishes brought in by the tidal currents into enclosed areas provided with sluice gates, where they are allowed to grow from three to nine months before harvesting. The total production of fish and prawns from aquaculture practices is about 10,000 tonnes. The potential in this sector appears to be most promising and by the year 2000 the annual production of 4 million tonnes (2.5 from coastal and 1.5 from fresh water areas) appears to be in this reach. World aqua culture production expected to reach 70 million tonnes by the year 2000 and this is supposed to come largely from developing countries (Table 2.1).

Sea-Weeds

Sea-weeds are one of the important living resources exploited by man for food, animal food, fertilizers and for chemical and pharmaceutical products. The total maritime algae yield of the world has been estimated as 172,000 tonnes per year, of this India's contribution only 1.09 per cent of the total (Table 2.2). The demand for agrophytes and alginophytes by industry in India and abroad is increasing very rapidly. Unfortunately India has not yet fully utilised its sea weeds resources (Table 2.3). The sea weed based industries in India came into existence only during the past twenty years. Before the advent of an indigenous sea-weed industry about 200 tonnes of dried sea-weed used to be exported from India annually,

TABLE 2.1: Total Brakish Water Areas Available in different Maritime States of India Together with the areas under Cultivation and Annual Production[9]

State	Potential Areas Available for Culture (ha)	Areas currently under culture (ha)	Annual Production (tonnes)
Gujarat	376,000	88	NA
Maharashtra	81,000	—	—
Goa	19,000	—	—
Karnataka	8,000	4,800	1,000
Kerala	243,000	5,700	6,000
Tamil Nadu	80,000	—	—
Andhra Pradesh	200,000	—	—
Orissa	299,000	—	—
West Bengal	405,000	20,000	—
	1,711,000	30,588	10,000

TABLE 2.2: Annual Yield of Marine Algae From Different Countries as Per Cent of Total World Proudction[10]

Sl. No.	Countries	Annual Yield %
1.	Japan	62.26
2.	Republic of Korea	10.63
3.	Norway	9.27
4.	Canada	4.37
5.	Mexico	3.52
6.	United Kingdom	3.09
7.	India	1.09
8.	France	1.09
9.	Spain	1.08
10.	West African Countries	0.45
11.	Madagaskar	0.04
12.	United States	0.02
		100.00

Coral Reefs

Coral reefs are among the most biological productive taxonomically important living communities. While their massive occurrence provides much needed protection for coastline from waves, the biological productivity yields a multitude of fauna and flora dependent on the coral reef ecosystem. The communities also form the main attraction of skin diving, underwater photography, sport fishing and shell collecting. The only reefs that would probably survive would be on the atolls of Laccadives and on some of the islands of Andaman and Nicobar.

TABLE 2.3: Total Seaweed Landings in India

Area	Annual Yield in Tonnes (Fresh Weight)
Gujarat	7,000
Maharashtra	6,000
Goa	1,000
Tamil Nadu	5,000

After discussing living biotic resources of world oceans it becomes essential to know the exploitation of abiotic resources of oceans of the world in next chapter.

NOTES

1. Majid Husain, "Human and Economical Geography" p. 1, National Council of Education Research and Training Reprint 1986.
2. *Ibid.* p. 2.
3. Readers Digest Great World Atlas. p. 120.
4. *Ibid.* p. 120.
5. Article of S.Z. Qasim, "A Technological Forecast of Ocean Research and Development in India", p. 297. Published in "Strategic Digest", March 1985. Institute for Defence Studies and Analysis.
6. Majid Hussain, "Human and Economic Geography" p. 14, National Council of Education Research and Training Reprint 1986.
7. *Ibid.* p. 15.
8. *Ibid.* p. 15-16.
9. S.Z. Qasim's Article on "A Technological Forecast of Ocean Research and Development in India", p. 299 - Published in Strategic Digest, March 1985 - Institute for Defence Studies and Analysis.
10. *Ibid.* p. 300.

3

GEOGRAPHY OF ABIOTIC RESOURCES AND MINERALS OF OCEANS

समुद्रः पीत कौशेय वाससी समुपाहरत ।
वरुणः स्मत्रं बैजयन्तीम मधुनामत षटपदाम ॥[1]

The ocean gave yellow coloured clothes to Lord and pearls garland (*Baijayanti Mala*) on which the buzzbee (Bhawmwara) were buzzing.[1]

Introduction

During Rigvedic Era 1200 B.C. to 2000 B.C. the oceans were churned by the *Devtas* and *Rakshyasas* and 14 gems (*Ratnas*) named below were found.

श्री, मणि, रम्भा, वारुणी, अमिय शंख गजराज ।
कल्पवृक्ष, शशि, धेनु, धनु धन्वन्तरि विष, बाज ॥

Goddess *Lakshi*, *Kaustubhmani*, *Rambha* Angel, Doctor *Dhanwantari*, Moon, *Erabat* Elephant, Poison, Nectar, Shark Bow, Horse, *Kamdhenu* Cow, Kalpatru, Snails, *Sura*, and Liquor were the 14 gems explored after churning of sea.

The Abiotic Resources

In this category all non-living minerals i.e. manganese, mica, copper, iron, silver, gold, sulphur, sand stone, tin, barites and tunguston are included. Before the investigator describes the reserves of polymattalic nodules existing in oceans of the world one should know the geology which forms the basis of these rocks and reserves.

Geology of Marine Sediments

The complex process of sediments accumulation is known as sedimentation or sedimentary genesis.[2]
Sediments particles may be classified according to origin as under:-

(a) Lithogenous (derived from rocks) Components are primarily mineral grains derived from soils formed on continent and transported to the ocean by rivers or wind. Volcanoes also contribute to these as volcanic ash.
(b) Biogenous (derived from organism) Components are skeletal remains such as shells, bones and teeth of marine organism. They contribute calcium carbonate, sillica or phosphate minerals to the sediments.
(c) Hydrogenous (derived from water) Components originate from inorganic chemical reactions in sea water or in the sediment itself. Manganese nodule is a typical example. Terrigenous sediments came under lithogenous variety. They consist of various elastic and clayey material. Chemogenous sediments, that are chemically precipitating from the sea water is grouped under hydrogenous. The biogenic sediments are also known as organogenous variety.[3]
(d) Volanogenic or eruptive sediments are obtained from products of submarine and above water volcanoes.
(e) Polygenic sediments like red oceanic clay are developed due to operation of numerous factors.

The Exploitation of Abiotic Minerals

The floor of the ocean is known to contain large quantities of minerals besides biotic resources. These minerals have much economic value. Some of the regions are found to be containing large deposits of minerals such as monazite, ilmenite, phosphorite etc. The continental shelf areas as revealed through geological sediments have sufficient qualities of reserves of oil, natural gas, coal, gold, alluvial diamonds, tin, sulphur etc. In the world of fast depleting raw materials, the ocean floor holds out bright promise of abundant resources of minerals for mankind.[4]

The non-living minerals resources can be grouped and described as under. They are discussed in the subsequent paras:-

(a) Resources from beach and sea-bed in shallow waters.
(b) Resources buried in deep sediments.
(c) Resources found in underlying rock.
(d) Resources obtained from sea water.

Resources From Beach and Sea Bed in Shallow Waters (Continental Margin)

Beach deposits, concentration of heavy minerals in beaches (placer deposits) submerged beaches, surfacial deposits of continental shelf and phosphorites mostly occur in the continental margin.[5] The wind and water carry materials such as diamonds, gold, sand and mud.[6] Placers can be found for all types of minerals, the most common being sand, gravel and shells. Appendix 'F' shows different types of minerals currently obtained from placer deposits in various parts of the world (Fig. 3.1). After oil and gas sand and gravel constitute the second most valuable minerals extracted from sea, placer deposits. Other materials mined in large quantities from the ocean are barite (barium sulphate) used in drilling mud which is mined off the Alaskan coast, iron sands and cassiterite (tinoxide). Placer containing diamond, gold and other high value minerals are also commercially exploited in various parts of the world. The black sand occur in many localities along the Indian coast.

Deposits on the West coast are largely concentrated as high grade beach and low grade dune deposits, extending from Kanyakumari to the Maharashtra coast with interruptions in between. These deposits mainly contain ilmenite, rutile, zircan and monazite with varying proportions of magnetite and garnet.[7] Minerals placers are also reported on the coastal tracks of many countries in the world. However, economically exploitable deposits are known only from the beaches and the shelf of Australia and India (rutile Zircan), India and Brazil (Monazite and ilmanite), Norway and Japan (Magnetite), Malaysia (Walfra-mite and Cassiterite), South Africa (Chromite and precious diamonds), Alaska, Canada, U.S.S.R. and South Africa (native gold and platinum). The U.S.S.R. and United States (ilmenite and rutile). Apart from these major occurrences of placer deposits, minor deposits have also been found off Mozambique, Senegal, Indonesia, Korean Peninsula, Thailand and Sri Lanka.[8] Many important metallic placer minerals are magnetite, ilmenite, chormite, rutile, monazite, cassiterite, zircon, welfaramite, columlite, tantalite

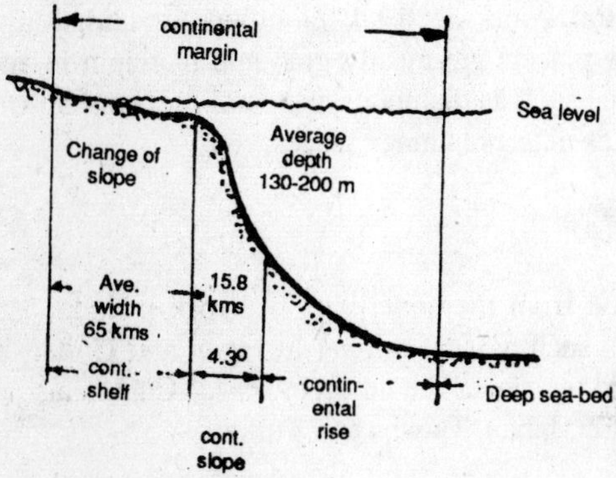

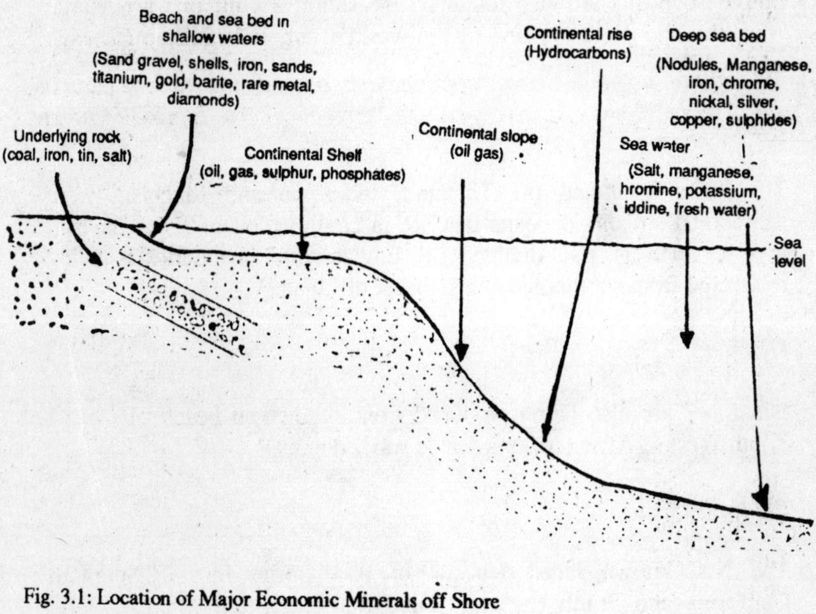

Fig. 3.1: Location of Major Economic Minerals off Shore

and cinnabar. These minerals along with garnet and such other non-metallic placers get themselves dislodged from metamorphic and igneous rock under normal geological processes and find place on the beaches and littoral zones as black or coloured sand. Many native elements occur as placers are mostly gold and platinum in addition to silver.[9] Many gems such as diamonds and semi-precious stones such as garnet also occur as placers in many places.

Coal

Minerals recovered from the underlying bed-rock includes land based mining operations such as coal mining in Japan and United Kingdom that have extended out under the adjacent seas.[10] Coal is also mined in Canada, U.K., Chile, Japan, Taiwan and Turkey.

Bauxite

Bauxite deposits are also found on the shore of Gulf of Carpentaria, Australia and presently one of the world's largest bauxite mine exist near sea.[11]

Tin

Tin is being mined off Thailand, Indonesia and Malaysia from submerged streams deposits that are in continuation of rocks on land and sea. Tin is also dredged off England and magnetite is being recovered from submerged beach placer of Japan.[12]

Diamond

Diamonds are also being produced from submerged beach placer off South West of Africa in comparable water depths.[13]

Phosphorite

The best known large deposits of phosphorite lie off Southern California and South Eastern United States, Eastern Mexico, Peru, Chile and Union of South Africa. Additional large deposits probably also occur off North West Africa and Western Australia. Although sea floor phosphorite accumulations are large, they seldom contain sufficient phosphate to meet the cut off grade for fertilizer production which is the chief use of phosphorite.

Resources buried in Deep Sediments

These sediments have generally high content of organic material because of life processes in the overlying water during their deposition and this, combined with their favourable thickness, is favourable for oil and gas accumulation. Sand and gravel are other valuable minerals extracted from the buried deep sediments.

Oil and gas are of organic origin. The organic material buried in marine beds underwent changes to produce natural hydrocarbons and these subsequently moved into reservoir rocks and accumulated to form oil pools. The slow decomposition of plant and animal organisms in the absence of oxygen is responsible for the formation of petroleum hydrocarbons. The lower plankatonia organisms such as diatoms and algae that thrive abundantly near the surface of the sea are the most probable and important. So oil deposits could be either of animal origin or plant origin. During the conversion of organic matter into petroleum, and natural gas, methane is also formed predominantly. The greater part of the oil so far produced is from structural 'Highs' viz anticlines, domes and salt domes though oil pools have resulted from faulting and igneous intrusions.[14]

At present off shore petroleum comes from areas with 150 km of the coast and from depths of about 100 m. However, in the near future, petroleum will be obtained from depths of 2000 m and more. Due to increase of highly sophisticated technology and greater demand, more wells of petroleum will be drilled in deep seas.

Exploration of oil is presently under way on continental shelves of 75 countries, 28 of which are producing or about to produce sub sea oil and gas. World wide offshore production has now grown to 17 per cent of the total world daily output of 5.5 million tonnes of oil and offshore reserve amount to 21 per cent of the total reserve of 60 billion tonnes. In addition more than one hundred 300 meter core holes have been drilled by deep water drilling ships in water depths upward of 1500 metres on the upper continental slope of Gulf of Mexico and the Western Atlantic Ocean. Many similar core holes are currently being drilled in water depths up to 6,000 metres by the drilling vessel Glammer Challenger as part of United States Deep Sea Drilling Project. Oil and sulphur reserves were found.[15] Currently over 20 per cent of the total petroleum production of the world is obtained from offshore sources. It has been estimated that by 1990 the figure will be raised to 50 per cent. The current value of offshore oil production is over 50 billion dollars a year and it is over 90 per cent of the total value of sea linked mineral resource extraction.[16]

The Indian continental shelf covering about 320,000 sq. km can be broadly sub-divided into ten sedimentary basins. Of these preliminary geophysical investigations on the Kutch, the Bengal, the Cauvery, the Mahanadi, the Godavari basins and a few others have already been undertaken. The areas in the West Coast near Ratnagiri and Bassein have been proved to be economical and production in Bombay High region has already been started.

The oil sector's performance in sixth plan in India has been impressive which is described in Table 3.1.[17]

TABLE 3.1: Oil Sector's Performance in Sixth Plan

	Activities	Anticipated Achievements	Remarks
A ON-SHORE			
1.	Survey (party years) Seismic	200	
2.	Exploratory Drilling		
	(a) Rig years	166	
	(b) No of wells	310	
3.	Development Drilling		
	(a) Rig year	73	
	(b) No wells	371	
4.	Oil Productions (MMT)	40.42	
5.	Gas Production (MMMM^3)	10.488	
6.	L.P.G. Production (000 Tonnes)	73.2	
B OFF-SHORE			
1.	Surveys Seismic (std. line km × 1000)	122.9	
2.	Exploratory Drilling		
	(a) Rig years	29.0	
	(b) No of wells	101.0	
3.	Development Drilling		
	(a) Rig years	23.5	
	(b) No of wells	247	
4.	Oil production (MMT)	63.1	
5.	Gas production (MMMM^3)	5.148	
6.	L.P.G. Productions (000 Tonnes)	682.0*	

* includes NGL.

The growth of oil production in India in 1970-75 was only 36.27 million tonnes which grew to 103.26 million metric tonnes in the 6th plan period representing a three fold growth in a decade. The performance on the production of gas shows similarly that from a production of 8,299 billion cubic metres in 1970-75, the production during 1980-85 was 23,820 representing a three-fold increase during the last decade. During 1980-81 we drilled 119 wells annually, 1984-85 we drilled over 240 wells. The crude oil production in 1980-81 was 10.51 million cubic metric tonnes which in 1984-85 it is expected to be over 29 million tonnes representing an almost three fold increase. The gas production has also shown a similar increase from 2,358 billion cubic metres to 4.32 billion cubic metres. During 6th plan period a number of major off shore projects were commissioned by ONGC.[18] These are:-

(a) BHN Feb 81	—	The first process platform complex off shore India.
(b) BHS Aug 82	—	The largest off shore complex in India today.
(c) RATNA Nov 82	—	The first satellite platform outside Bombay High Field.
(d) WIN Mar 84	—	The first water Injection platform Bombay High North Field.
(e) HEERA Nov 84	—	The second satellite field outside Bombay High Field.
(f) HOO Mar 85	—	The Northernmost process Complex in the Bombay High Field for Development of the L-11 horizon.

As estimated by ONGC in Bay of Bengal at a distance 8.5 km South West of Annalpuram in Andhra Pradesh on 01 January 1988 at the depth of 2035 mtrs. with a flow of 400 barrels of oil and 25000 cubic metres of gas per day will be produced. The Krishna-Cauvery bases, the Andaman and Nicobar island should hold good hydrocarbon deposits estimated as 1400 million tonnes. Pranita and Godavari basins in Andhra Pradesh and Pari basin in Tamil Nadu and Cauvery basin in Tamil Nadu produced 7.15 lacs barrels on 01 Jan 1988 (See further efforts at Appendices G and H for crude oil in India).

Polymetallic Nodules

The floor of the Pacific Indian and Atlantic Ocean is spread with polymetallic nodules which have a concentration of valuable metals such as copper, nickel, cobalt and zinc. The highest concentration of these nodules has been found at depths ranging from 3500 metres to 6000 metres and covering a total area of 46 million sq.km in the world oceans. Of this area, some 23 million sq. km are in the Pacific Ocean, 15 million sq. km. in the Indian Ocean and 8 million sq. km in the Atlantic Ocean. Total resources of these nodules are estimated to be more than 1.7 trillion tonnes, of which the Pacific Ocean has 1.5 trillion tonnes, the Indian Ocean has 0.15 trillion tonnes and the Atlantic some 0.05 trillion tonnes.[19] Polymetallic nodules are renewable resources for they keep forming at the rate of 10 million tonnes every year. these nodules contain manganese, iron, nickel, copper, cobalt, lead, malybdenum, cadmium, vanadium, titanium, silver, gold and other metals.[20] The density of a group of nodules ranges between 2.07 — 3.07. Manganese nodules exhibit complex layering and variety of nodular structures (See Appendices I and J).[21] The chemical composition of these nodules is described in Table 3.2.

TABLE 3.2: Metal Contents of Polymetallic Nodules (Percentage Weight)

Sample No.	Mn	Fe	Co	Ni	Cn
1.	9.50	17.00	0.16	0.13	0.07
2.	25.60	4.10	0.16	1.57	1.28
3.	18.63	19.64	0.48	0.31	0.13
4.	42.30	2.47	0.17	0.26	0.15

The information on their occurrence and distribution is largely derived by sediments samples and deep sea photographs. It vary from 3 to 18 percentage in Table 3.3.

The details of area covered by nodules in three oceans of the world is given in Table 3.4.

Indian interest in the nodules goes back to 1977 when National Institute of Oceanography at Goa started a project in this field. In January 1981, the research vessel 'Gaveshani' collected several samples of nodules from the Arabian Sea for the first time. India has identified two mining sites in Indian Ocean. It has also earned status of "Pioneer Investor" because of its work in the field. The third UN

Conference has conferred this status. India shares this distinction with USSR, France, Japan and four multinational consortia. The central Indian basin has to be surveyed to demarcate the site for mining. The size of one pioneer area should not exceed 150000 sq. km.

TABLE 3.3: Percentage Occurrence of Polymetallic Nodules in Four Different Regions of the World Oceans[22]

Area	Total Stations	Occurrence of Nodules	Percentage
North Pacific Ocean (0-50°N, 90°W 110°E)	16546	3004	18
South Pacific Ocean (0.70°S 70°W 110°E)	9077	1118	12
Indian Ocean (70°S-20°N 110-30°E)	5406	486	9
Atlantic Ocean (70°S - 70°N)	19178	654	3
All Oceans	50207	5262	10

TABLE 3.4: Area covered by Nodules in the Three Oceans (Million km^2)

Oceans	Area Covered	Prime Areas
Pacific	23	5.20
Atlantic	8	0.85
Indian	15	0.50

Resource found in the Underlying Rock

The principal materials currently extracted from the off-shore bed rock are coal, iron, tin, bauxite and salt. In the case of coal, Japan produces over 40 per cent of its total off-shore, while other countries extracting appreciable amounts of coal from under the sea are Australia, Turkey, Great Britain, Canada and Chile. Recently off-shore artificial islands have been built to provide entrance to sub bottom mines. Table 3.5 shows the total value of marine mineral production and percentage of the world.

Resources obtained from Sea Water

All eroded material ultimately reaches the sea and most of it settles on the bottom as sediments. Some out of this are soluble and some remain as it is. The undermentioned important minerals are found in water:

Salt

Salt is widely produced on coasts by pumping sea water into shallow ponds where it is evaporated by the sun. This method is followed especially in low humidity, hot, desert areas. Salt is used for human consumption, a raw material to obtain chlorine for use in plastic, rubber, caustic soda.

Magnesium

Magnesium is required for a wide variety of chemical and pharmaceutical products. Sea water contains about 0.13 per cent of magnesium which can be extracted by a precipitation with calcium hydroxide. This process is currently the main source of magnesium world wide. Bromine, Potassium and Iodine are also elements extracted commercially from sea water.

TABLE 3.5: Total Estimated Output, Value of Production of Minerals and Percentage of the World, 1977

Mineral	Output 000's* (Metric Tons or** noted)	Value of Production (Millions of US Dollars)	% of World Total
Petroleum	$10.5x16^{6}$^	50,000	20.2
Gas	$18.3x10^{9}$^^	10,000	7.2
Coal	34,036	—	1.4
Sulphur	1,500	36	2.9
Iron Ore	30	0.5	1.6
Sand and Gravel	85,900	189.4	1.0
Calcium Carbonate	16,290	95.2	1.0
Tin	13.7	190	6.1
Barite	378	0.1	1.0
Salt Brine	151	0.05	NA

* Barrels per day
** Cubic feet per day

Water

Probably the more immediate importance and the biggest possible source in many desertic countries is the recovery of fresh water from the sea. Large quantities are recovered by Kuwait, Abu Dhabi, Saudi Arabia, Malta, the Canary Islands, Bahamas, U.S.A., Mexico, Southern U.S.S.R. and Netherlands.

NON-LIVING RESOURCES

Fresh water From the Sea

The supply of world fresh water is dwindling rapidly in comparison to the growth of population. There are many areas in India where potable water is in short supply. People drink saline water which causes flvarosis — a disease leading painful symptoms of bone deformity. The biggest possible source in many desertic countries is the recovery of fresh water from the sea. Large quantities of fresh water are recovered from sea by Kuwait, Abudhabi, Saudi Arabia, Malta, the Canary Islands, Bahamas, the U.S.A., Mexico, Southern U.S.S.R. and Netherlands.

The following processes of disalination are being employed to generate fresh water from saline water of the sea.

Solar Stills

Solar stills are ideal for small coastal village as they run on a non-expendable energy source. There is no problem of power failure. A solar still of 5,000 liters per day capacity has been installed in Avnia villages in Gujarat where 500 families obtain drinking water from this source.

Electrodialysis

This technique employsion selective members for the disalination of brakish water. Electrodialysis in more economical for salinities below 5000 ppm. The energy cost of the process is directly proportional to the salinity thus beyond 5000 ppm. This process is uneconomical and slow.

Reverse Osmosis

It is the most widely used desalination technique. In this process, suitable osmatic membuanes are used which reject salts and allow the water to pass through, several plants with capacities of 1000 to 15,000 litres have been set up in Indian villages, to supply drinking potable water.

Drugs from Sea Water

Marine planktons or plants and animals are utilised for making safe and effective useful drugs by the pharmaceutical agencies of the 100 or more organisms that have been screened so far, forty-two have given promising results. The important feature in the anti-fertility properties of ten marine organisms — Recently Prostaglandins, which play a major role in controlling biological reproduction have been isolated from seaweed Gracilarish by Australian researchers studies in India indicate that all those species of marine algae that exhibit antifertility properties may also contain prostaglandins. Researches in India in this field are continued on the cultures of marine bacteria, fungi, yeast etc. Some fruitful results will accrue for the production of bioactive substances in near future.

Marine Chemicals

It is difficult to use the sea water for chemical elements. However, of the sixty elements present in sea water only six are recovered commercially. These are sodium and chlorine in the form of common salt, magnesium in the form of its compound, bromine, calcium sulphate (gypsum). It is possible to recover potassium from bittens (mother liquor from salt extraction). Efforts are being made to recover many useful elements, commercially, normally iodine, Uranium and gold from sea water.

The Energy from the Waves

This energy is available and extracted more from water of temperature climates than waters of tropical seas. Several littoral countries, are actually engaged in devising systems to convert energy from the waves into electrical energy. Japan have conducted many sea trials using ship based system for energy. The Norwegians have commissioned a 0.5 MW commercial wave energy plant in 1985. This system leaves a clam pool of water behind the system which in later can utilised for small harbours for small craft and for aqua culture in addition, fishing can be done by using smaller crafts. It helps coastal, protection and better coastal transportation by way of using faster crafts.

This system consists of putting a suitable barrier in the sea and allow water to rush through reversible turbines. The technology for the construction of such a barrier is well established, and know how already available for the construction of harbour and docks can be used for this purpose. The France tidal power plant of 240 MW capacity is in operation since 1966.

OCEAN IN DEVELOPMENT OF ENERGY

Oceans are helping in production of energy which is considered to have a long time potential of satisfying much of the World's need for energy. The chief sources of energy from the ocean are:–

(a) Ocean Thermal (OTEC)
(b) Ocean Tides.
(c) Ocean Waves.
(d) Ocean Salinity Gradients.
(e) Currents.

Among all the only economical, and potential sources of energy, harnessing from Ocean is available from Ocean Thermal, waves and tides. Other forces are very weak.

The Ocean Thermal (OTEC)*

The OTEC system used the temperature differential that exists between the warm upper layers and the cold depth (between 700 and 1000 m). The oceans of tropical and sub-tropical regions offer several tides of Ocean Thermal. This type of energy resources is perennial and hence provides the land electricity unlike land based solar energy converts the temperature differential of about 20°C can be used in a rankine cycle to develop power. Though overall efficiency is small (about 2.5%) available is enormous because of the virtually infinite volume of the ocean. The estimated World OTEC potential is 10,000 MW and off

* Article Prof Dr. ING, V.S. Raju "Ocean Resources and Ocean Engineering Aspects" P-1-9, Ocean Engineering Centre, I.I.T., Madras-600036.

the Indian coast is 3,000 MW. This process will also prove to be more useful, attractive to provide energy for deep sea mining of various minerals. The suitable side for India are Lakshadweep and Andaman Islands.

To conclude historically Indians had the know how to extract minerals from the oceans in ancient period. The complex geology of sedimentation is found beneath the ocean floor. The deposits lithogenous, biogenous hydrogenous, volcanogenic and polygenic sedimentation from the basis of rocks and mineral resources found in oceans. Various minerals are exploited. These are diamond, gold, sand, corals, shells, rutile, zircon, monazite, ilmanite, walframite, diamond, chromite, coal, bauxite, oil, polymetallic nodules, tin, salt, magnesium, water and ocean energy. Oceans will be the greatest treasure house of minerals and sources of energy in future.

NOTES

1. *Srimad Bhagwat Mahapuran Ashtam Skand Sloka* 35.
2. Article of Dr. S.P. Subramanian "Geological Oceanography" p. III-6, Ocean Engineering Centre, ITI, Madras.
3. *Ibid.* p. III.7.
4. Pamphlet "Polymetallic Nodules India's Work on Resources for Future". Department of Ocean Development. Government of India, New Delhi.
5. Article of Dr. S.P. Subramanian, *op. cit.* p. III-8.
6. Article of Prof. Dr. V.S. Raju, "Introductory Lecture on Ocean Resources and Ocean Engineering Aspects" p. 1-3. Ocean Engineering Centre, I.I.T. Madras.
7. Article of Mr S.Z. Qasim "A Technological Forecast of Ocean Research and Development in India" p. 302. Published in Strategic Digest March 1985, Institute for Defence Studies and Analyses.
8. *Ibid.* p. 302.
9. Article of Dr. S.P. Subramanian, *op. cit.* p. III-8.
10. Edited by Richard C. Vetter, "Oceanography the Last Frontier". pp. 160-161, Voice of America Forum Series.
11. *Ibid.* pp. 161-163.
12. *Ibid.* pp. 160-163.
13. *Ibid.*
14. Article of Dr. S.P. Subramanian, *op. cit.* p. III-10.
15. Edited by Richard C. Vetter, *op. cit.* pp. 155-157 — 1974.
16. Article of Prof. V.S. Raju, *op. cit.* p. I-6.
17. Article of Dr. A.K. Malhotra, "Off Shore Activities related to Oil and Gas in India". pp. 1, 2. International Consultant, Formerly Member of O.N.G.C., Bombay.
18. *Ibid.* p. 3.
19. Pamphlet, "Polymetallic Nodules. India's work on the resources for the future". Dept. of Ocean Development, Government of India, New Delhi.

20. *Ibid.*
21. Article Dr. S.P. Subramanian, *op. cit*. pp. III-12.
22. Article S.Z. Qasim, *op. cit*. p. 303.
23. Kent, Peter, E., "Minerals from the Marine Environment". Edward Arnold Ltd, London, 1980.

4

EXPLOITATION OF OCEAN MINERALS BY DEVELOPED COUNTRIES AND SCIENTIFIC TECHNOLOGY USED

Early Explorations

SEA IS the earth's water reservoir and provides the least expensive form of transportation. The early voyages of discovery by Columbus, Magellan and Cook extended man's geographical knowledge of the oceans. The first serious attempt to explore the ocean body and ocean floor was made by the British Challenger Expedition (1872-1876). This global voyage marked the beginning of the modern oceanography. The oceanographic institutions began to develop in 1920s and 1930s.

The Modern Era

After World War II the real progress in oceanography began to develop. U.S.A., U.S.S.R., Great Britain, France, Japan, West Germany and other developed countries started researches in exploitation of ocean floor, the ocean plains and ocean depths. Relief maps of the ocean floor such as Bruee Heegah and Marie Tharpe were published. The early work of Maurice Eiving and his colleagues at the Lamant Dohetry Geological Observation of Columbia University laid the ground work of modern knowledge of ocean sediments. A great advance in ocean explorations came with the development of deep diving vehicles equipped with Portholes. In the beginning the nodules were solely of scientific interest. But now ocean engineers and some mining men have become interested in sea bed resources. These resources have become subject of international organisation the U.N. In 1977 a United Nations programme called the International Decade of Ocean Exploration (IDOE) was begun. The project has involved more than 35 countries in all branches of oceanography. In project

MODE (Mid Ocean Dynamics Experiment, 1971) and POLYMODE (1973) the U.S.A., Great Britain and U.S.S.R. cooperated in studies of short-lived oceanic currents called rings. In the GEOSECS (Geo-Chemical Ocean Section Study) programme many nations are cooperating in sampling and analysing of sea water in order to gain a better understanding of ocean circulation. In the FAMOUS (French American Mid Ocean Under Sea Study) programme of 1973-74 three submersibles were used to study a section of mid Atlantic ridge in great detail. Biologics are involved in several projects of IDOF.[1] The international phase of ocean drilling (IPOD) began in 1972. It is an out growth of the Deep Sea Drilling Project (DSOP) and also of earlier unsuccessful attempt to drill through the earth's oceanic crust into the mantle. The Glammer Challengers drill has penetrated almost a mile into the ocean floor in water more than three miles (5 km) deep. Its belt is hollow at the centre, thus allowing the recovery of rocks and sediments samples. Drill cores have been taken from more than 400 sites around the world giving ocean geologist access to exciting new material for minerals, its history and evolution of the earth.[2]

Before the investigator describes modern technique for exploitation of deep sea mining resources one should understand that by deep sea bed is meant the floor of the ocean beyond the limits of national political jurisdiction either under existing definitions or under the proposed 200 miles economic zone. The deep sea bed lies at an average depth of about 4000 m.[3]

Technology Applied for Mining of Nodules

The floor of the Pacific, Indian and Atlantic Oceans is spread with polymetallic nodules which have a concentration of valuable metals such as copper, nickel, cobalt and zinc. The highest concentration of these nodules has been found at depths ranging from 3500 metres to 6000 metres and covering a total area of 46 million sq km in the world oceans.[4]

Several methods based on dredging techniques have been devised to lift the nodules from the deep sea. The following important lifting systems are discussed in some detail:

(a) The deep sea dredge.
(b) Continuous bucket line system.
(c) Air lift system.

(d) Hydraulic lift system.

The Deep Sea Wire Line Drag Dredge

This is most probably the least complicated method of mining sea floor nodules. This drag dredging operation would require two floating vessels. The dredging ship 2000 ton ocean going tug type vessel with an installed 2000 hp winch for dredging to about 4000 m, should have excellent manoeuvrability. The ship should be designed so that main generation can serve as a power source for the dredge winch. During mining operations only a small fraction of the ship's power output would be needed for propulsion. The second vessel, an ocean going barge, would be large enough to store about 5000 tonnes of mined and cleaned nodules. This barge would carry equipment to separate the nodules from the debris recovered with the nodules and propulsion unit to enable it to follow the dredge ship during mining operations. It would have facilities for rapid transfer of mined material to waiting transport vessel unless the barge were to be used as the transport vessel itself. The dredge bucket system should be as large as possible that would allow safe working conditions at sea. On a 2000 ton ship the largest bucket practicable would be about 6 × 4 × 1 m. The weight of this bucket would be about 3 tonnes and its load capacity would be about 13 tonnes haul assuming about 65 per cent filling efficiency. If 25 per cent of the material dredged from sea bed is assumed to be gangue such as shark's teeth, whale ear bones and fine sediments etc., this bucket should cover about 10 tonnes of nodules per haul. The bucket will be allowed to drop at a free fall velocity to the sea floor, which, with proper hydrodynamic design, the dredge bucket should be as such as 200 m per minute. A sonic device would be attached to the bucket so that the operator would know when the bucket touched the sea floor. The bucket would be dragged across the sea floor until it was filled with nodules and then would be retrieved.[5]

Continuous Bucket Line System

This system (Fig. 4.1) can be considered as an improvement over drag dredge because CLB system has less idling time and also continuous output. A number of buckets of approximately 80 × 40 × 25 cm in size are arranged in the form of continuous loop. This loop is driven by a tracking machine on board the ship. The length of wire in the loop is taken sufficiently long so that buckets have sufficient time to be dragged on the floor of the mining site. The major disadvantages of the system are the tangling of bucket line and non-availability of sufficient information on the ship bridge about the behaviour of the buckets on the ocean floor of the mining site. To overcome the above difficulties model tests and field tests in shallow water are in progress at various places. The first deep sea mining using CLB system was done in the Pacific by a Japanese ship of 1865 tonnes DWT and 75 metres long. The major features of this CLB system are:

Depth of mining	—	2800 metres.
Length of rope	—	8400 metres.
Spacing of Buckets	—	25 metres.
Diametre of rope	—	40 mm
Shape of Bucket	—	Rectangular (80 × 40 23 × cm)
Rope speed	—	0.3 to 0.5 m/sec.
Rope tension	—	3 tonnes.

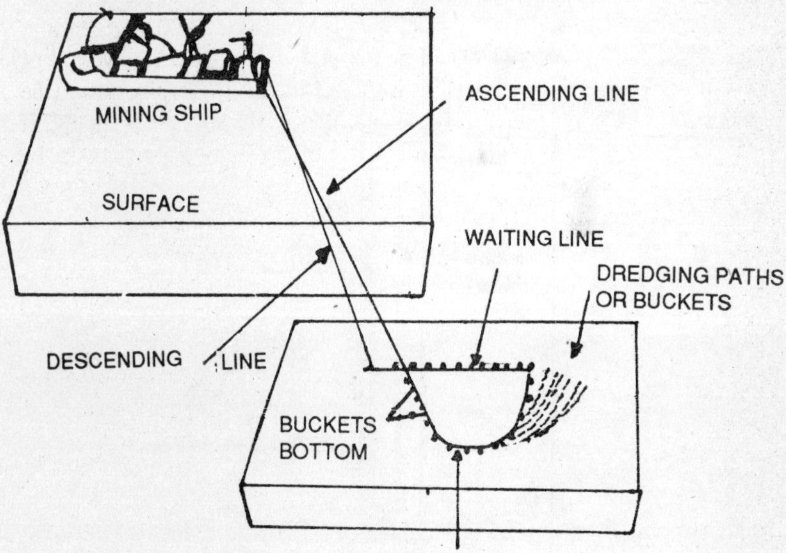

FIG. 4.1: CLB Mining System

Air Lift System

In this device, a long pipe with both ends open is taken and one end is lowered close to the nodules on the sea floor. The pipe rises to the top and other end is kept just about the sea water level. Compressed air is injected near the bottom of the pipe is lessened forcing the fluid column to rise in the pipe line. If the top of the pipe is not too far about the surface of the water the air water mixture will over flow it. This causes water to enter the pipe at the bottom of the pipe and to rise vertically. If there are nodules and other solid substances near the bottom end of the pipe they will be drawn into the pipe. The important condition is that the inlet velocity of water into the pipe at the bottom end should be large enough to accelerate the nodules into movement. For the continuous vertical movement of the nodules, the velocity of water in the bottom of pipe (below the air inlet point) should be larger than the velocity of sinking of the nodules. Below the air inlet section, there is a vertical flow of two phases namely water and nodules while above the air inlet section, there is a three phase flow with water air and nodules. As this three phase flow moves upwards, the air expands due to fall in hydrostatic pressure and, therefore, all the three phases are accelerated (Fig. 4.2).

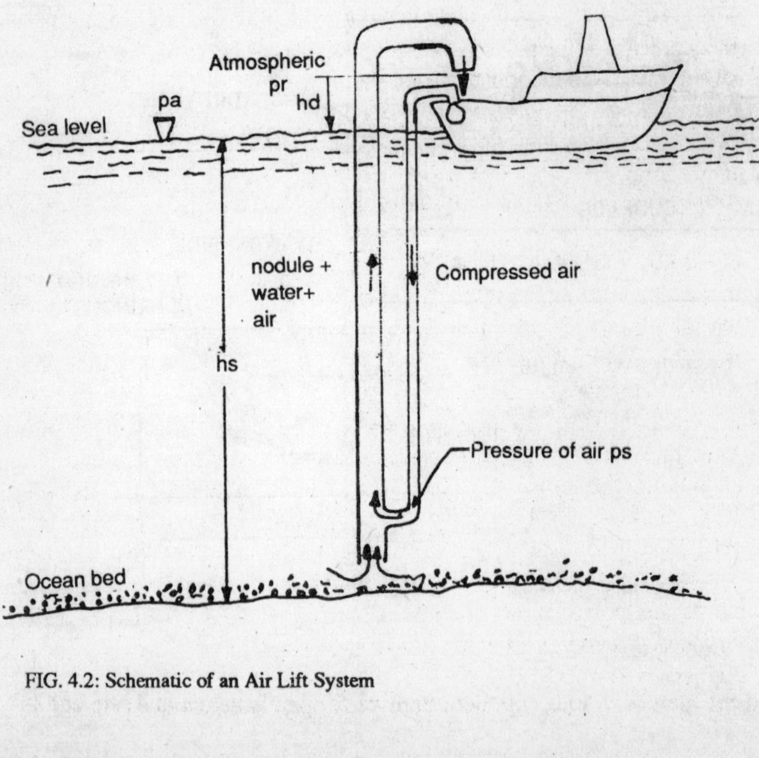

FIG. 4.2: Schematic of an Air Lift System

Hydraulic Lifting Device

The hydraulic lifting devices are being proposed for large scale mining of nodules from the floor of oceans of the world. Extensive model and site tests have been conducted. The nodules are suspended in sea water and the mixture is pumped to the surface. The pumping system consists of several units of centrifugal pump and submersible motors placed at different intervals along the hydraulic line. Earlier proposals for hydraulic dredging system were supposed to work as giant vaccum cleaners, the pumps sucking water along with nodules from ocean bed. But the recent designs separate nodules from sediments and supply the nodule water slurry to the vertical riser pipes which have pumps at regular intervals. The collector could be of other passive or active design. Passive collectors require no power to operate other than towing of the collector across the sea floor. The passive collector may just have conical shape to concentrate the nodules at the bottom end of the riser cables. But the collection capacity of such a device is limited.[6]

Active collector concepts require external power for operation. These collectors could have either a technical or hydraulic system. The mechanical concepts generally involve the raking, scooping or sweeping of the nodules from the sea floor. These designs make extensive use of mechanical components like clutches, gear boxes, bearings, sprockets and chains. Since these components work in an atmosphere of deep sea water and fine sea bed sediments, lack of proper maintenance will lead to problems quickly. The hydraulic concepts, because of limited number of moving parts are expected to be simple and most reliable. This system consists of its own travelling mechanism driven by submersible motor. A submersible pump flushes the nodules and feeds them to the riser pipe through a flexible hose for hydraulic lifting. Figure 4.3 shows the system of hydraulic lifting along with the collector.

To conclude, the early discoveries of oceans by Columbus, Megellan and Cook extended geographical knowledge of the oceans and continents. It was the expedition of British Challenger (1872-1976 AD) marked the beginning of modern oceanography. Developed countries such as U.S.A., U.K., Japan, France and West Germany started exploitation of ocean resources after Second World War. For extracting polymetallic nodules concentrated in the beds of Pacific, Atlantic and Indian Oceans, following four devices are being used in common:

(a) The deep sea dredge system
(b) Continuous bucket line system
(d) Air life system
(d) Hydraulic lift system.

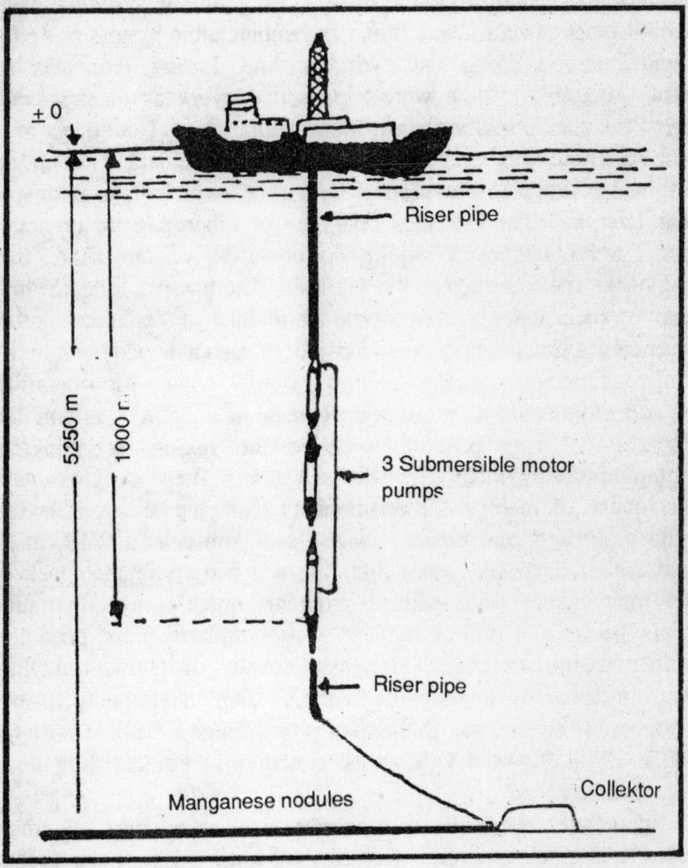

FIG. 4.3 : Schematic of Hydraulic Lift System

Minerals exploited from the nodules will strengthen the output of national resources.

THE ASPECT OF OCEAN ENGINEERING

Ocean Engineering is a new multidisciplinary branch of technology development among other things for exploration and exploitation of the immense resources of the oceans. In addition to their resources, the seas and oceans are used for transportation, navigation, recreation and as a dumping ground for wastes etc.

This engineering task is more expensive than to land activities. Moreover, the hazards of work are greater and the management of operations more demanding. As a rule the farther a structure lies, from shore or the greater distance from the source of supply, the more prodigious will be the cost of any marine operation. The special problems pertaining to ocean engineering include designs and construction to withstand very high pressure and hydro-elastic forces, due to flow and wave action especially on slender structures, estimation of wave and current forces in a particular location, transport of polyphase mixtures via submarine pipes, design and analysis of submarine pipe-lines and offshore structure, pipe laying techniques under different sea conditions, control of under water machinery, under water surveying, oceanographic measurement, analysis of data etc. The problems can be broadly divided as follows:

Management of Coastal Zonal Affairs

The large coastal zone of all the littoral states of India is being used for the development of natural and artificial ports and harbours, fisheries, beach resort, land reclamation, location of shore-based industrial complex, human resettlement, agriculture, disposal of wastes, arrangement of tourist resort etc. These states are confronted with problems, like pollution in estuaries and nearshore water. The increase in offshore activities for oil and gas extraction, fisheries, mining of seabed minerals, extraction of energy, shipping and marine transportation poses special problems to provide facilities for offshore drilling alongwith natural environment.

Ports and Harbours

During this era of scientific development increase in the volume of sea borne traffic in petroleum, its products, minerals like tin, coal, fertilizers, foodgrain has enhanced the development of new harbours and expansion of existing harbours and construction of new ports. India has to adopt new method of large cargo shipping. To avoid congestion, diversion to new ports is essential.

Control of Coastal Erosion

Coastal erosion has become the hindrance to cargo ships and frigates. Various ports of the western and eastern coast of India are confronting with severe erosion. The waves are the most affecting factor causing continuous soil erosion. Besides waves, tides, currents, beach material are also affecting the erosion. Human being are also affecting natural coastal margins by construction or destruction of some layer of the coast. Engineering measures designed to coastal erosion and stabilise the eroding shore can be described as under. It includes constructing structures such as sea walls, revertments, groynes, break-waters and jetties to reduce or prevent the wave energy from reaching the erodible material on the shore. Artificial filling of shore line with or without structure, like groynes, reduce the rate of loss of entrapped material.

Check of Pollution of the Coastal Zone

India has a very long coastline of more than 6000 kms. The increase in heavy industrialisation and population pressure the estuaries, basins and near shore waters of continental shelf are being polluted as a result of the disposal of waste materials, sewage discharge from large cities harbours, agriculture waste and dumpy of oil pollution are also causing great harm to the coastal zone oil pollution arises from tanker's accidents and deblasting operations and tank washing refinery off events, municipal and industrial wastes, losses from pipe lines and offshore production platforms. To drive solutions to control pollution of waste material the oil cleaning equipment is to be installed. Suitable oil booms and skimmers are to be developed or obtained.

Production of Oil from High Seas

It needs too many engineering efforts to dug and install a platform for oil wells in high seas.

NOTES

1. *Merit Students Encyclopaedia*. p. 4.
2. *Ibid*. p. 4.
3. Article of Dr. M. Ravindran, "Deep Sea Mining" of Ocean Engineering Centre, I.I.T., Madras.
4. Pamphlet, "Polymetallic Nodules India's Work on the Resources for the Future". Department of Ocean Development, Government of India, New Delhi.
5. Article of Dr. M. Ravindran *op. cit*.
6. *Ibid*. p. VII-9.

5

EFFORTS OF INDIA IN EXPLOITATION OF MINERALS AND RESOURCES AT ANTARCTICA

"Man marks the earth with ruin his control stops with the shore"

—Byron

AS TIME passes on, man is looking more and more towards the natural factors and adjacent sea for raw materials and food. Many countries in the world including India are exploiting the sea, particularly its floor, for oil, gas and other minerals.[1] Besides this a new regime of the oceans has come into being in December 1982 when at Jamaica, more than 100 countries affixed their signature on to a new convention to govern the wealth and uses of the seas. These new laws hold an immense promise of development particularly for the developing nations in as much as their rights and sharing of ocean wealth are protected.[2] A Marine Research and Development Fund (MRDF) has been created in the Department of Ocean Development as a part of its effort to encourage meaningful ocean related activities in different institutions of the country including private bodies. The support of a variety of organisations i.e. universities, scientific institutes and industrial units has been asked and used. The subjects considered for support under the fund include physical and chemical oceanography, marine biology, marine geology, marine geophysics, ocean engineering, marine ecology, meteorology, marine instrumentation etc.[3] Department of Ocean Development was created on 24 July, 1981 directly under the late Prime Minister, Mrs Indira Gandhi. The department is to vigorously pursue research and development activities in all branches of ocean science.

Before the investigator describes about Antarctica Expeditions the various projects undertaken by the Department of Ocean Development in exploitation of minerals living as well as non-living are described in this paper. The systematic surveys and explorations of marine living and non-living resources in the Exclusive Economic Zone and the continental shelf of the country is being carried out by two vessels of the department namely FORV Sagar Sampada and RRV Sagar Kanya. FORV Sagar Sampada extended the survey programme on living resources in relation to environmental conditions. Designing suitable fishing gear for the exploitation of both conventional and non-conventional variety of fishes, crustaceans and ceppalopods for commercial exploitation by the private industry in due course formed an important objective of courses under taken during the year under report. In a period of 283 days of her 17 research cruises, she covered 94,146 line km.[4] Scientist from a number of research institutes and other organisations including CMEFRI, CIFT, FSI, Annamalai University Madras, Kerala Agriculture University, I.I.T., Bombay, Vikram University, Berhampur University, Zoological Survey of India, Central Agriculture Research Institute, Port Blair, Andhra University etc. have participated in these research cruises.

India is one of the leading maritime countries and has a deep interest in the sea due to richness of resources in the ocean. The Department of Ocean Development has undertaken various programmes i.e. exploration of Antarctica, non-living resources, under water technology, acquisition of new research vessels and ship management, marine pollution research, remote sensing, living resources, manpower development, data collection and dissemination, energy from the sea, desalination, legal regime, use of marine research and development fund. India has been accepted as pioneer investor for ocean researches (Fig. 5.1).

Brief History of Antarctica

There are five claimant of ownership i.e. U.K., Newzealand, Australia, Norway and France. Later U.S.S.R. Belgium, Japan, Sweden, Poland and South Africa joined these explorations. In 1959 an Antarctica Treaty existing for 30 years was signed by 12 nations at Washington. Provisions of this treaty opened the Antarctica to unrestricted scientific explorations in spirit of international cooperation.[5]

Expeditions

The first Indian expedition team to Antarctica landed on 9 January

1982. This was a national venture for establishing a place for Indian scientist in the polar regions. India could have sent 8 expedition teams to this continent. This first expedition had set up a solar powered automatic weather recording station at Dakshin Gangotri on 9 January, 1982. This station was programmed to activate meteorological sensors and record the data on wind speed, wind direction, humidity and temperature. The second expedition reached on 28 December, 1982,

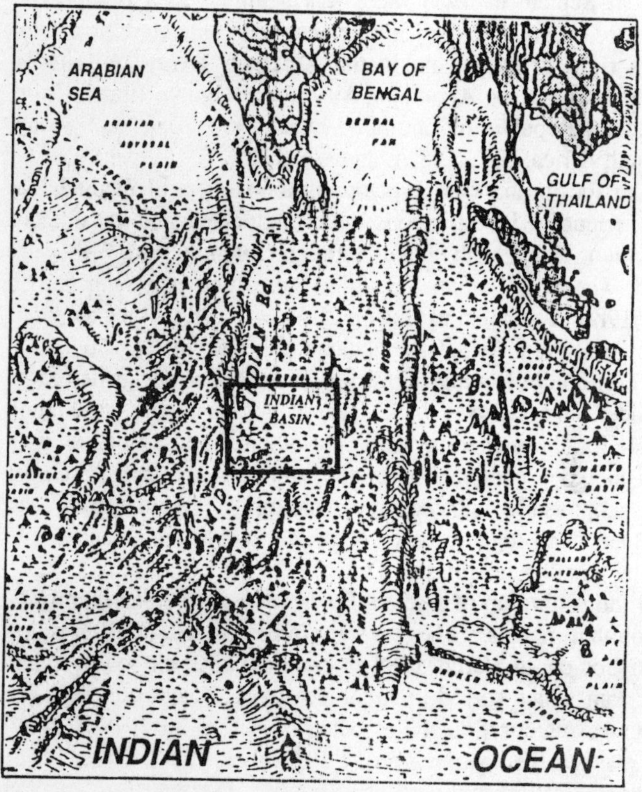

Location of the Indian Pioneer area

Fig. 5.1: Location of Indian Pioneer Area

selected the second site for building a permanent station and identified a suitable site for an air strip. Besides continuing scientific studies initiated by the earlier expedition, the team in third expedition established on 27 December, 1983 a direct communication link between the base camp in Antarctica and India and also between base camp and mobile parties on land and on the ship. The third expedition built a permanent two storeyed manned station which can accommodate 15 scientists. India had already accepted to the Antarctic Treaty in August 1983, thereby joining earlier signatory countries of the treaty. In September 1983 India was admitted as a consultative party to the treaty. India is the 15th member state to acquire this status and with a right to participate in the Antarctic Treaty consultative meetings, the other being Argentina, Australia, Belgium, Brazil, Chile, France, Federal Republic of Germany, Japan, New Zealand, Norway, Poland, South Africa, U.K., U.S.A. and U.S.S.R.[6]

Later the fourth and fifth expeditions infrastructural facilities for continuing scientific investigations were augmented and satisfactory communication links with India through satellite and HF were established. The sixth Indian Scientific Expedition was launched in November 1986. It returned to Goa on 22 March, 1987. The expedition led by Dr. A.H. Parulekar of the National Institute of Oceanography. Shri M.L. Parimoo of Geological Survey of India was designated as its deputy leader and Colonel V.S. Iyer of Indian Army as Station Commander for the wintering team. The scientific tasks of the sixth expedition included continuation of the studies in the thrust areas of geology, geophysics, meteorology, biology, oceanography and geomagnetism. The logistic task comprised of the maintenance and repair of the existing facilities of Dakshin Gangotri station and Maitri Camp and the setting up of a scientific camp at the Peter mannmassif in the wohlthat ranges.[7] The seventh Indian Antarctic Expedition returned to Goa on 26 March, 1988 and brought back the fourth Indian wintering team which stayed in Antarctica from December 1986. The eighth Indian Antarctic Expedition sailed from Goa on 29 November, 1988 and reached the continent on 24 December, 1988. A post office was inaugurated at Dakshin Gangotri. Republic Day was celebrated at the New Maitree Station on 26 January, 1989.

The Approach for Exploitation of Resources

It is quite difficult to predict separately about the oceanic and land resources of Antarctica. Yet there is a greater scope of getting living and non living resources.

Living Resources

Antarctic waters are among the most biologically productive provinces on earth. The abundant krill in the world's largest, microscopic fungi, algae, lichens and mosses make up the vegetable life while krill, fishes, 300 plankton, penguins, sea birds, crabe eater seals, dolphins, odontocate, mysticte, weddel seal, whale, insects, mites, circumpolar ross seals, leopard seal, elephant geols, ticks and lice constitute the animal life in the region.[8] The shrimp like krill is the most important link in the food chain of the Antarctic waters. Quantitative estimates of phytoplankton shaves dominance of a few species such as fragilaria cylindrus and conscinodiscus during winter months and as photoperiod increased the number of phytoplankton spices also increase. Marine mammals are also available in Antarctic waters.

Non-living Resources

Antarctica is, therefore, believed to have many minerals resources such as coal, iron, copper, lead, uranium, oil and gas which are found in the separate land masses today. Indian expeditions will exploit some of these resources of minerals in near future. Exploration and drilling is required to extract petroleum out of sedimentary rocks lying over Antarctica continent.

To summarise the department of ocean development is trying to extract treasure resources of Indian and Antarctic oceans since its inception in 1982. Two ships Sagar Sampada and Sagar Kanya are doing investigations to know and explore living and non living resources in the seas. India has become a member of 15 members countries who are doing scientific researches in Antarctica continent. The continent and its oceans have rich scope of living and non living resources. By now India had sent 8th expedition teams to Antarctica from 1982 to 1988.

NOTES

1. Pamphlet "Polymetallic Nodules India's Work on the Resources for the Future". Department of Ocean Development, Government of India, New Delhi.

2. Pamphlet "Activities of the Department of Ocean Development in the Field of Ocean Science and Technology", p. 1, Department of Ocean Development, Government of India, New Delhi.
3. Pamphlet, "Marine Science Promotion of Research and Development Activities". Department of Ocean Development, Government of India, New Delhi.
4. Annual Report 1988-89, p. 3, Department of Ocean Development, Government of India, New Delhi.
5. *Encyclopaedia of Britannica*, Vol. II, p. 6.
6. Pamphlet "India and the Antarctic Scenario", Department of Ocean Development, Government of India, New Delhi.
7. Annual Report 1987-88 P. 27, Department of Ocean Development, Government of India, New Delhi.
8. *Ibid.*

6

OCEAN AND MINERAL TRADE ROUTES

"Who ever controls the Indian Ocean dominates Asia. This Ocean is the key to the seven seas. In the twenty-first century the destiny of the World will be decided on its waters".

—*Alfred Mahan*

GEOGRAPHICALLY THE above statement may not be true due to existence of the littoral countries, which are under developed or semi-developed both industrially and from the point military. But it is altogether true that oceans are the keys to the existence, development and progress of a nation — Indian Ocean too have maintained its significance since times immemorial in exploitation of minerals.

As exploitation of minerals from the sea has started with the advent of scientific technology and exertion of pressure of population upon the resources of the country having sea coast. It will be difficult to describe trade and shipping of minerals of a country exploitated from the land or beds of the ocean Countries. Nations which have coastal frontiers or are littoral have many more benefits from the ocean besides biotic and abiotic minerals. Ocean influence the climatic conditions of the littoral countries and connect them to international foreign trade. The existence of ocean waters, deeps, bays and straits is now-a-days no problem for national cargo ships and naval fleets and frigates.

Locationally as well as strategically countries which are land locked are not advanced countries except a few. Mangolia, Nepal are land locked countries in Asia, Czechoslovakia, Austria, Hungary are the countries located in Europe having no approach to oceans while — Central Africa, Chad, Zambia, Rhodesia, Mali, Botswana land, Niger, Uganda are the countries in Africa which have no link to oceanic coast. While at the same time Paraguay, and Bolivia are the two countries which are land locked in South America. These countries for trade and commerce have to accept under compulsion the policies of countries where from they want to establish friendly relations. Some times neighbouring countries stop essential commodities which cause famine, inflation and internal insecurity among land locked countries. With the great advance in technology ocean — going ships have developed carrying capacity far larger than railways, roadways and air craft. By increasing the dimensions of Ocean carriers, large volumes of world trade can be transported at a very cheap cost. The introduction of refrigerated chambers for transporting, biotic and perishable goods such as meat, fruits, vegetables have improved the efficiency of sea trade. Now a days modern liners and cargo ships are equipped with radar, wireless and navigable aids and are driven by powerful steam, diesel and nuclear powers. Winds, storms, hurricanes and weather are no more a problem.

The Importance of the Oceans for Mineral Trade

The important ocean trade routes are described as under:

The Indian Ocean Route

The area of the Indian ocean is rich in several minerals, raw-materials and marine foods has been regarded as one of the world's major storehouse of natural resources. This region particularly accounts for 80.7 per cent of world extraction of gold, 56.6 per cent of tin, 28.5 per cent of nickel, 18.5 per cent of bauxite, 12.5 per cent zinc and 77.3 per cent natural rubber.[1] The importance of the Indian Ocean has tremendously enhanced due to availability of largest seams of oil bearing rocks — on its coast.

The West Asian countries known as gulf countries possess the potential deposits of oil and gas. Important countries producing oil are Iraq, Iran, Saudi Arabia, Amman, Qatar, Kuwait, Bahrain, United Arab Emirates, Yema Arab Republic, Southern Yeman. The oil reserves of the West Asian countries account approximately 50 per cent of worlds trade which pass through Gulf and Indian Ocean. Plenty of warm water fishes are found on the coasts of Pakistan, India, Bangladesh, Ceylon, Burma, Malaya, Indochina and Indonesia. At least 90 per cent of worlds rubber comes from Malaysia, Indonesia, Cambodia, Thailand and Sri Lanka and transported to Japan, Australia, U.S.A. and European countries through Indian Ocean. All the jute products of Bangladesh and Thailand are transported to developed countries through Indian Ocean. Advanced nations have invested their money in various corporations and companies to explore oil prospects from the littoral base of the undeveloped countries. The Indian ocean has become a very important area for navigation and marine trade and two-thirds of oil products are loaded and unloaded in the ports of this region. About 60000 vessels pass through the lanes of Indian Ocean. A large number of vessels pass through Suez Canal, Cape of Good Hope and Malacca strait. About 1000 vessels pass through Babel mandib strait (at the mouth of Red Sea) every year. It has been intimated that from 80 to 90 vessels full of oil and minerals pass through strait of Hormuz every day. It has also been estimated that about 140 vessels pass from Malacca Strait every day. Nearly 20 million barrels and 800,000 barrels pass every day and hour respectively from strait of Harmuz. The power who succeeds in controlling these strategic entries i.e. Suez Canal, straits of Babel Mandib, Harmuz, Malacca and the route of Cape of Good Hope will be able to rule and dominate predominantly in South West, south and south east Asia.

The Suez Canal Route

The Suez Canal (Fig. 6.1) opened in 1869, is one of the great achievement of man in the field of ocean transportations, navigation and commercial trade. It linked the continent of Europe with West Asia and South Asia and far East. It provided a shorter route between European countries and Asian countries. It had been and will remain the 'Life line of trade' for both developed and undeveloped countries because of oil supplied from the middle East, tropical raw materials and food products from the Asian countries went through the Suez. India imports 16 million tons of oil, comes from Gulf countries through Arabian ocean sea route. Manufactured products of heavy machines of textile, Iron, and defence tools are imported by way of the Suez to Bombay, Karachi, Colombo, Singapore, Hongkong and Australian ports. The European countries import oil, rubber, tin, jute and cotton through this route. The length of the canal is 162.5 km, having depth 10 metres, width 60 — 65.5 metres and tonnage limit 2000 tons.

The Cape of Good Hope Route

This route connects Europe, countries of West coast of Africa to with the countries of South East Asia and Far East. Through this route minerals like tin, bauxite, chromite and copper, mica, manganese and raw materials like rubber, jute, sugar, tea, coffee from South East Asia, South Asia, South Africa are transported to Europe. It helps the nations at the closure of Suez Canal route. It links Atlantic ocean with that of Indian Ocean in Southern hemisphere.

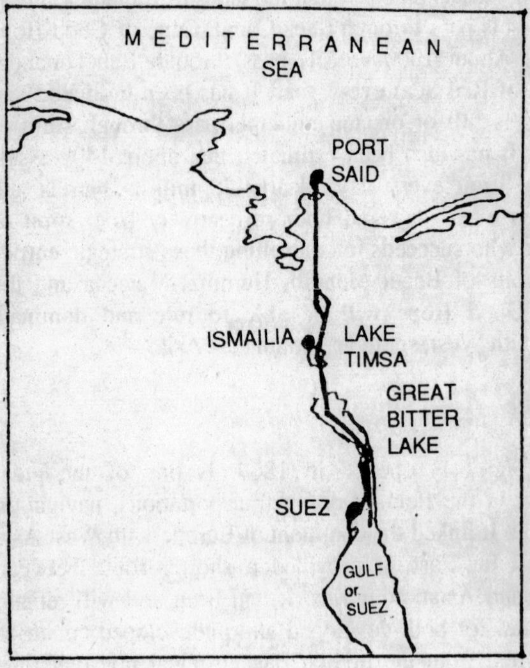

FIG. 6.1: Suez Canal

Date of Construction	:	1869
Length	:	162.5 kms
Depth	:	10 mtrs
Width	:	60-65.5 mtrs
Tonnage Limit	:	20000 Tonnes

The Pacific and Atlantic Ocean Route

The Panama Canal

The Panama (Fig. 6.2) canal was constructed by U.S.A. in 1914 at altitude 25.9 metres to 0.5.1. It has a length of 81.6 kms and width 91.4 — 304.8 metres with a depth of 12.5 — 25.9 metres. It separates North America to South America and connects the two great oceans — the Pacific in the south and Atlantic in the North. It eliminated the long diversion of hazardous voyage round the stormy cape Horn. It is the 'Gateway to the Pacific' which has benefited countries on both the sides of Atlantic and Pacific sea coasts. It has also facilitated trade of Brazilian rubber to Japan and Mexican oil to India. Raw materials and manufactured products can be easily transported from the Western Coast of Atlantic ocean to the eastern coast of Pacific ocean. The construction of the Panama canal has saved more than 4000 kilometres from Auckland to Newyork instead of the route through Cape Horn. The only drawback of this route is that every ship has to pass three gates which have different water levels. The entry to this route is through clearance of the control rooms which are responsible for making the water level equal by filling water from lakes through tube wells or some other scientific devices.

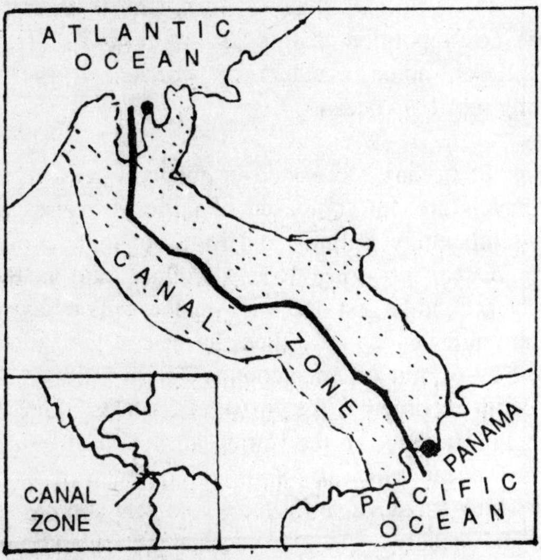

FIG. 6.2: Panama Canal

Date of Completed	:	1914
Altitute Maximum	:	25.9 mtrs ASI
Length	:	81.6 kms
Width	:	91.4-304.8 mtrs
Depth	:	12.5-25.9 mtrs

NOTE

1. Lt Col (Retd) Dr. Bhupinder Singh, 'Indian Ocean and Regional Security", 1st Edn. 1983, B.C. Publishers, 47 Ajit Nagar opposite Fatima Convent Patiala 147001, Punjab (India).

7

CONCLUSIONS AND SUGGESTIONS

पुरोधसां च मुख्यं विद्धि पार्थ वृहस्पतिम ।[1]
सेनानीनामहं स्कन्दः सरसामस्मि सागरः ॥

— Srimad Bhagwat *Geeta* Ch. 10, *sloka* 24

[O'Arjun see my cosmopolition nature, I am *Vrahaspati* (Jupiter) among priests, among leaders of worriers I am Kartikeya and among seas I am ocean.[1]]

THE SOURCE of water is the oceans. Oceans cover about 71 per cent of the earth's surface. Oceans are not composed of uniform masses of water. They differ in temperature, salinity and forms of marine life. Among oceans Pacific Ocean covering 165,384,000 sq. km is the largest in the world. The second largest ocean is Atlantic, which has an area of 32 million square miles which is 10 times the size of the United States. It separates the New World America continent from Europe and Africa. The greatest depth recorded is the MilWU-Kee Deep (30250 feet), off the island of Purerto Rico in the Caribbean Sea. The Indian Ocean is the third largest ocean which lies South of continent of Asia. This is the only ocean which is navigable for the whole year and has no effect of cold climate and icebergs. The other oceans are Antarctic as well as Arctic Oceans consisting of 11.7 million square km and 1,40,89,600 sq. km of areas respectively. The ocean floor is divided into continental margin, shelf, slope and rise. With the advent of scientific technology oceans have become a source of food energy, raw material, commercial and shipping. The coastal boundary ranges from 3 nautical miles to 200 nautical miles for various countries.

A resource has been defined as any means of attaining given ends. There are two important types of resource of ocean, the biotic and abiotic. The living things, i.e. forests, agriculture, animals wild and domestic, birds, reptiles and fishes are included among biotic. They can continue to reproduce and regenerate their population or species in existing environmental conditions. The non-living minerals i.e. coal, petroleum, natural gas, iron, aluminium, gold, silver, platinum, manganese ore, cobalt and nickel are included in series of abiotic resources. The principle fishing grounds of the world are North-West Atlantic and North-East Atlantic. India has a long coast line, yet only about 8 lakhs tonnes of marine fish are caught annually. Fish production increased to 17 million tonnes. Herring, mackerel, prawns, catfish, mullets, ponfret, Indian salmon, jawfish and flounders are important fish found in Indian Ocean. Other resources are seaweed and coral reefs.

During *Rigvedic* Era, oceans were churned by the *Devatas* and *Rakshyasas* and 14 gems (*Ratnas*) were found. Some of them were biotic and others were abiotic. The process of sedimentation on sea beds is complex.[2] These deposits being lithogenous (derived from rocks), biogenous (derived from organism), hydrogenous (derived from water), volcanogenic or (eruptive) and polygenic (red oceanic clay) form the basis of various polymetallic nodules and minerals. In the world of fast depleting raw materials, the ocean floor holds out bright promise of abundant resources of minerals for mankind.[3]

The non-living or abiotic mineral resources of ocean can be described into four categories as under:

(a) Resources from beach and sea bed in shallow waters.
(b) Resources burried in deep sediments.
(c) Resources found in underlying rocks.
(d) Resources obtained from sea water.

Among the first type placer deposits like diamond, gold, sand and mud are the minerals found.[4] Iron, barium, sulphate and cassiterite (tinoxide), ilmenite, rutile, zircon, magnetite and garnet are also found in shallow waters and beaches.[5] Australia and India are exploiting rutile and zircon, monazite and ilmenite, Norway and Japan are exploiting magnetile while Malaysia walframite and cassiterite. South Africa is getting chromite and diamond. Canada, U.S.S.R. are exploiting gold and platinum. Japan, United Kingdom, Chile, Taiwan, Turkey and Canada are recovering and mining coal from the oceans. Australia is extracting bauxite. Indonesia, Malaysia and Thailand are exploiting tin. South Africa is producing diamonds from water depths of seas. Phosphorite is extracted by Sn California, Wn Mexico, Peru and Chile from sea bed. In the second group oil and gas are extracted from continental shelves of 75 countries. World-wide offshore production of oil and gas has grown to 17 per cent of the total world daily output. Gulf of Mexico and Indian continental shelf hold rich prospect of oil and gas. The floor of Pacific, Atlantic and Indian Oceans hold rich reserves of polymetallic nodules full of manifold minerals. In third group coal, iron, tin, bauxite and salts are found from adjacent seas of Australia, Great Britain, Canada, Chile and Turkey. Among the fourth category salt, magnesium, water and oceanic energy are obtained.

The early voyages of discovery by Columbus, Magellan and Cook extended man's geographical knowledge of the oceans. The British Challenger's expedition (1872-1876) revolutioned the knowledge of ocean body and floor. Oceanographic institutions began to develop in 1920s and 1930s. Later in modern era U.S.A., U.S.S.R., Great Britain, France, Japan, West Germany started researches in exploitation of oc[illegible] floors, plain and deeps. The deep sea bed is the floor of the deep oceans beyond the limits of national political jurisdiction of 200 miles. The deep sea bed lies at an average depth of about 4000 m.[6]

The advanced countries are using following methods under new technology:

(a) The deep sea dredge requires two floating vessels.
(b) The continuous bucket system combines a number of buckets approximately 80×40×25 cm and arranged in the form of continuous loop.
(c) The air lift system is a device in which a long pipe with both ends open is taken and one end is lowered close to the nodules on sea floor. Compressed air is injected near the bottom of the pipe.
(d) The hydraulic lifting devices are being proposed for large scale mining of nodules from the floor of oceans of the world. The nodules are suspended in sea water and the mixture is pumped to the surface.

Many countries in the world including India are exploiting the sea, particularly its floor for oil, gas and other minerals.[7] During December 1982 more than 100 countries at Jamaica affixed their signatures on to a new convention to govern the wealth and uses of the seas. The right of developing countries for the use of ocean is safely protected under this convention. The Department of Ocean Development was created on 24 July, 1981 directly under the late Prime Minister, Mrs. Indira Gandhi. A Marine Research and Development Fund (MRDF) has been created to encourage meaningful ocean researches. Assistance of various institutions, universities, private bodies and industrial units have been asked for it. The subject considered for study and research are physical, chemical oceanography, marine biology, geology, geophysics, ecology, ocean engineering and marine instrumentation.[8]

There are two vessels under the Department of Ocean Development namely the FORV Sagar Sampada and RRV Sagar Kanya which are used to explore resources of Indian Ocean. India is one of the leading maritime country and has done 8 expeditions to explore resources of Antarctica. In 1959 an Antarctica Treaty existing for 30 years was signed by 12 nations at Washington. India had acceded to this treaty and got the right as 15th member of unrestricted scientific explorations at Antarctica. The first Indian team landed on Antarctica on 9 January, 1982. The second expedition reached the site on 28 December, 1982 and identified suitable site for airstrip. The third expedition established on 27 December, 1982 a camp named Dakshina Gangotri with direct communication link. Like others the 8th Indian Expedition reached Antarctica on 24 December, 1988 and established a post office. Antarctica waters are among the most biological productive provinces on earth and have largest microscopic fungi, algae, lichens and mosses, krill, fishes, planktons, penguin, sea birds and crabe eater, seals, dolphins, seal and whales. Among non-living minerals like coal, iron, copper, lead, uranium and oil and gas seem to be in sufficient quantity. Exploration is a hazardous task to the scientist. As regards oceans and mineral trade routes the important routes are mainly Suez Route, Panama Route, Cape and Good Hope, Cape of Horn and Mallaca strait. The important resources and minerals i.e. fishes, rubber, cotton, wheat, cocoa, tunguston, crude petroleum, manganese, mica and machines made of minerals are transported from these routes from one country to another country. Now-a-days ocean transportation is much cheaper in comparison to airways and land transportation.

Ocean Policy Statement of the Government of India

The oceans are known to be our last frontier. Our long coast and the sense of adventure of our ancients fostered a great maritime tradition. The Indian Ocean which washes our shores provides opportunities which need to be utilized. For success in ocean development, the entire nation should be permeated by the spirit of enterprise and the desire to explore the frontiers of knowledge. Our experience in other fields of scientific endeavour will help our efforts in ocean development. What is necessary is a policy and structure to facilitate a dynamic thrust keeping in view the developments in other parts of the world.

The adoption, by an overwhelming majority of nations of the convention of the UN Conference on the Law of the Sea has established a new international order for the oceans. This extends the economic jurisdiction of coastal states to an area ranging to 360 km from the coastline. According to this regime, about 2 million square kilometres of area, or very nearly two-thirds of the land mass, has right to utilize living and non-living resources vests with the nation. Besides, India has been recognised as a "Pioneer Investor" in ocean mining. This gives India the exclusive right to operate in an area of up to 150,000 square kilometres in the high seas for the recovery and processing of polymetallic nodules.

For ages, the sea has enabled our people to sail to near and distant lands and has been a source of livelihood to large numbers of people. Even now Indian public and private enterprises do use ocean resources. The country is producing significant quantities of fish and hydrocarbons from the sea and much scientific work has been done in collecting basic knowledge and information about the sea and sea bed and in surveying, charting and exploiting it. Progress has also been made in construction and development of offshore structures.

The vastness, complexity and uncertainty of the ocean environment calls for a coordinated, centralised and highly sophisticated development response. This should be based on adequate knowledge of marine space (sea-bed, water and air columns included) as a fundamental prerequisite to the control, management and utilization of the rich and varied natural resources available in the sea. In addition to basic knowledge to determine the potentialities inherent in the Indian sea-space, we have to develop appropriate technologies to harness these resources. A supporting infrastructure has to be built. Effective system of management and control of the entire set up are also necessary.

We need to map living resources, prepare an inventory of commercially exploitable fauna and to map and assess the availability of minerals from the deep sea. The supporting infrastructure and incentives required are research vessels of different types, manpower, well-laid out programme of resources exploitation, advanced technology and everything necessary to promote the growth of ocean technology. In the management sector, the high seas and the exclusive economic zone up to 360 kilometres have to be looked into for the exploitation of the wealth occurring therein.

The main thrust should be on the optional utilization of living resources like fish and sea weeds, exploitation of non-living resources such as hydrocarbons and heavy placer deposits, harnessing of renewable resources of ocean energy from waves, temperature differences in the water column, tidal heights and salinity gradients and the collection and processing of polymetallic nodules from the deep sea.

Marine development is linked with scientific and technological achievements in other areas. Hence while we develop basic marine science and technology, i.e. technology for marine environment, our technological advances have to be geared to the utilization and preservation of the marine environment. The extension of national frontiers by an area of 2 million square kilometres of ocean space and the consequent access to new sources of energy, minerals and food, requires great strides in ocean engineering, specially in tasks related to structures, materials, instrumentation, submersibles and systems of propulsion of ships. The exploitation of natural food resources such as fish and sea weeds, and the generation of additional food resources by cultivation, also need scientific methods of aquaculture and mariculture. To survey and predict the ocean environment, the main tasks necessary are sea floor mapping, charting, geodesy, ocean dynamics, currents, waves, cyclones, marine fauna, chemistry and physics of the oceans and sea-bed mineral mapping, delineation and assessment. Research in all these arms must examine the various processes and their origins so as to have a fundamental understanding, ensuring predictive capabilities. Marine science and technology has also to look beyond the current state-of-the-art to achieve major technological breakthrough in future.

Besides research and development in basic sciences, we should survey the deeper parts of the ocean. Similarly in the deep sea, detailed survey and sampling in the regions of EEZ and the adjacent ocean will be necessary to locate and evaluate the rich and economically viable deposits of polymetallic nodules, heavy metals, fossil placers and phosphorite deposits. The gathering of data from surveys should be coordinated and cost-effective system of integrated surveys established.

Much more needs to be done for the development of indigenous technology for the exploitation of fish from deeper waters. This also means setting up of infrastructure facilities and services to operate large-sized fishing vessels.

An important component of the development programme should be acquisition of technology. To be self-reliant, such technologies would have to be largely developed, tested and operated indigenously. Technologies relating to instrumentation, diving systems, position fixing and position maintenance, materials, development,, ocean data collecting devices are priority items. Several new technologies will have to be commercialized and made cost-effective.

Infrastructural support forms an essential prerequisite for ocean development. The varied infrastructure already available in the country will have to be appropriately augmented, and more particularly in basic supporting facilities like safety and rescue at sea, navigation chains, communication net-work, development of appropriate maps and charts etc. Infrastructural support for providing a complete and reliable information system through a net-work of data centres on marine resources, processing and marketing systems, advanced technologies and financial assistance would also be necessary. This requires a broadening and strengthening of available infrastructural facilities. Provision of adequate ports and harbours, ship-building and ship-repair facilities will be needed in addition to adequate skilled manpower in various sectors of development.

Surveillance and conservation of the marine environment and its resources call for an integrated legal framework and its concomitant enforcement. Several laws have already been promulgated regarding the maritime zone, fisheries etc. The Coast Guard Organisation looks after the enforcement aspects of several of these legislative measures. The coordinating mechanisms of the overall structure of legislation will have to be suitably strengthened under the aegis of the Department of Ocean Development.

In the light of this, we must have a data-base to coordinate efforts made by different agencies. This is all the more necessary because of the rapid growth of information in ocean science and technology. A centralized data system will be set up by the Department of Ocean Development with a proper mechanism for collection, collation and dissemination of information acquired both indigenously and from foreign sources.

The creation of a self-reliant technological base puts a heavy demand on fully trained personnel. The training of skilled manpower is to be adequately planned. Young scientists, technologists and engineers will be encouraged to participate in the programme of ocean development and steps will be taken to induce Indian scientists from within the country and abroad to participate in it.

Existing agencies will have to be appropriately strengthened to meet the demands of this growing challenge. The Department of ocean Development will function in conjunction with other concerned agencies as a focal point to promote institutional capability in areas where significant work is lacking. The complex programme that ocean development entails will requires well-designed management and institutional extension of the Department of Ocean Development with sufficient powers *vis-a-vis* other agencies to help proper and speedy ocean development, which enables India to be in the forefront of the International effort. This would also mean close cooperation with both developing and developed countries in a spirit of understanding of the concept that the oceans are a common heritage of humankind.

Suggestions

The following are the suggestions on the investigation study:

(a) The process of exploitation of minerals and resources in the oceans is an expensive project. Seeing the growing requirement of minerals for rapid industrial development of India, more funds be earmarked for further exploitation and researches to the Department of Ocean Development of India.

(b) Intensive efforts be adopted for the development of biotic resources of Indian Ocean so that the food problem of India is minimised.

(c) There is a greater need for exploitation of oil and natural gas as the entire continental shelf and high seas of Arabian Ocean and Bay of Bengal hold large reserves under oceanic seams and beds. Very little effort has been made in the direction.

(d) Exploitation of resources of Oceans of India has rich aspects for scientist for research and industrialisation. It will open new vistas of challenges to marine Chemists, Engineers and Geologists too. Very little work has been done by

Oceanographers of under developed and semi-developed nations like India, Pakistan, Ceylon, Burma, Indonesia, Malaysia, gulf countries and coastal countries of Eastern Africa.

(e) To exploit and explore these resources India should invent or use its own sophisticated technology which will lead the nation towards self sufficiency.

(f) There are manifold fields of ocean i.e. Physical, Chemical Oceanography, Marine biology, Geophysics, Ecology, Engineering, Meteorology and Instrumentation which need more researches and funds to meet the growing requirement of nations of the world.

(g) The right of Nations for economic zone has increased up to 200 nautical miles at Jaimaca Convention in December 1982.

(h) To meet the enormous demands of industry concentrated efforts be taken by the nations to exploit raw materials, biotic and abiotic minerals from the ocean.

(i) India's oil resources need exploitation in ocean due to concentration of oil reserves in North Eastern states and dependence on Gulf countries. Besides this the growing need of industries, automobiles and civil aviation are pressurising countries to explore possibilities of drilling oil from coastal areas.

(j) India has a coast line of about 6000 km. It has exclusive rights to operate in area up to 150000 sq. km in high seas which would be equivalent to 60 per cent of the total land surface of the nation. Twenty-five per cent of population of India live in coastal states. About 170 million people are directly dependent on the sea for their livelihood. India has to revise and organise at a greater pace its planning to harness biotic and abiotic minerals of Indian Ocean as well as offshore riches.

(k) The Nation has been granted exclusive right to mine deep sea beds from offshore regions of Arabian Ocean and Bay of Bengal. The country hopes to mine 3 million tonnes of polymetallic nodules a year within next decade continued.

(l) Indian ocean including Bay of Bengal and Arabian Ocean has an estimated 15 million tonnes of the polymetallic nodules. These nodules contain approximately 15,000 tonnes of nickel, 1200 tonnes of copper, 60,000 tonnes of manganese, 1500 tonnes of cobalt and undefined quantity of gold.

(m) Strategically India is located at the position where from almost all nations of the world can be united through Indian sea routes.

(n) The South Asian Republic Countries Western and Eastern countries of Asia will have to look for assistance, cooperation from India to explore the minerals of their bays, oceans and seas. There is a need of greater understanding among Gulf countries, coastal Nations of East Africa, Pakistan, Maldieve, Ceylon, Burma, Malaysia, Thailand, Kampuchia, Laos, Java Sumatra, Borneo, Philippines and Australia

(o) The rights of ocean frontiers be guarded properly without harassing the adjacent and weak countries.

(p) Problems confronting with regard to jurisdiction of the zone should be tackled amicably.

(q) This research will become the guideline to various students of marine sciences, marine geology, biology, oceanic researches, scientists, university scholars and marine engineers.

(r) The developing nations like India, Pakistan and Indonesia have to earmark more funds to extract ocean resources in sufficient quantity. There is no harm in getting an expert advise from Americans, Japanese, Canadians and British Scientists for this development.

NOTES

1. Ved Vyas, "*Mahabharat — Geeta*" — Chapter 10, Sloka 24.
2. Article of Dr. S.P. Subramanian, "Geological Oceanography", p. III 6, Ocean Engineering Centre, I.I.T., Madras.
3. Pamphlet, "Polymetallic Nodules India's Work on the Resources for the Future". Department of Ocean Development, Government of India, New Delhi.
4. Article of Prof. Dr. V.S. Raju, "Introductory Lecture on Ocean Resources and Ocean Engineering Aspects" p. 1-3 Ocean Engineering Centre, I.I.T., Madras.
5. Article of Mr. S.Z. Qasim, "A Technological Forecast of Ocean Research and Development in India", p. 302, Published in Strategic Digest, March 1985.
6. Article of Dr. M. Ravindran, "Deep Sea Mining". Ocean Engineering Centre, I.I.T., Madras.
7. Pamphlet, "Polymetallic Nodules India's Work on the Resources of the Future". Department of Ocean Development, Government of India, New Delhi.
8. Pamphlet, "Marine Science Promotion of Research and Development Activities", Department of Ocean Eng. Development, Government of India, New Delhi.

APPENDICES

APPENDIX A

SURVEY OF OCEAN AND SEAS

Tables below show the surface areas, average depths and volumes of three oceans (with and without their marginal seas) of some of the smaller seas and lastly of the whole world oceans:

Surface Area, Volume and Average Depth of Oceans and Seas

		Area (in million km)	Volume (in million km)	Average depth (m)
Atlantic ocean	(Without marginal seas)	82.40	324.6	3930
Pacific ocean	(Without marginal seas)	165.25	707.6	4280
Indian ocean	(Without marginal seas)	73.44	291.0	3960
Arctic ocean		14.09	17.0	1205
Mediterranean seas & Black sea		2.97	4.2	1430
Gulf of Mexico and Caribbean sea		4.32	9.6	2220
Australian central sea		8.14	9.9	1210
Hudson Bay		1.23	0.16	128
Baltic Sea		0.42	0.02	55
North Sea		0.57	0.05	94
English Channel		0.075	0.004	54
Irish Sea		0.10	0.006	60
Sea of Okhotsk		1.53	1.3	838
Bering Sea		2.27	3.3	1440
Atlantic Ocean	With marginal seas	106.46	354.7	3330
Pacific Ocean	With marginal seas	179.68	723.7	4030
Indian Ocean	With marginal seas	74.92	291.9	3900
		361.06	1370.3	3795

Source: Groen, P., The World Ocean – The waters of sea, p. 30, D. Van Nostrand Co. Ltd., London.

Area, Volume and Mean Depth of Oceans and Seas

Body	Area ($10^6 \times km^2$)	Volume ($10^4 \times km^3$)	Mean depth (m)
1. Atlantic Ocean (excluding adjacent seas)	82.441	323.613	3923
2. Pacific Ocean (excluding adjacent seas)	165.246	707.555	4282
3. Indian Ocean (excluding adjacent seas)	73.443	291.030	3963
4. All Oceans excluding adjacent seas	321.130	1322.198	4177
5. Arctic Mediterranean	14.090	16.980	1205
6. American Mediterranean	4.319	9.573	2216
7. Mediterranean sea	2.966	4.238	1429
8. Asiatic Mediterranean	8.143	9.873	1212
9. Large Mediterranean sea	29.518	40.464	1378
10. Baltic Sea	0.422	0.028	55
11. Hudson sea	1.232	0.158	128
12. Red Sea	0.438	0.215	491
13. Persian Gulf	0.239	0.006	25
14. Small Mediterranean seas	2.331	0.402	172
15. All Mediterranean Seas	31.849	41.066	1289
16. North Sea	0.575	0.054	94
17. English Channel	0.075	0.004	54
18. Irish Sea	0.103	0.006	60
19. Gulf of St Lawrence	0.238	0.030	127
20. Andaman Sea	0.798	0.694	870
21. Bering Sea	2.268	3.259	1437
22. Okhotsk Sea	1.528	1.179	858
23. Japan Sea	1.008	1.361	1350
24. East China Sea	1.249	0.235	188
25. Gulf of California	0.162	0.132	813
26. Bases Strait	0.075	0.005	70
27. Martinal Seas	8.079	7.059	874
28. All Adjacent Seas	39.928	48.125	1205
29. Atlantic Ocean (Including adjacent seas)	166.463	354.679	3332
30. Pacific Ocean (Including adjacent seas)	179.679	723.699	4028
31. Indian Ocean (Including adjacent seas)	74.917	291.945	3897
32. All Oceans (including adjacent)	361.059	1370.323	3795

APPENDIX B

OCEAN BOUNDARY AND CLAIMS OF THE TERRITORIAL SEA

Oceans as Boundary

Base

3 Nautical Miles from Low tide	4 Nautical miles (Scandinavian Countries (Norway, Sweden Finland)	5 miles Cambodia	6 miles zone marginal sea Spain, Portugal (Latin, American Countries)	6 & 8 miles Honduras

9 miles	10 miles	12 miles offshore (Russia, Pakistan, India)	250 miles limits (Chile, Peru, Ecuador, France Sansalvador)	Epicontinental sea (Latin, American)	Continental slope

National Resources of sea beds of continental shelf 180 meters or 100 fathom (America)	Ocean Deeps	High Seas

"Territorial Sea"

"Man marks the earth with ruin — his control stops with the shore."

—Byron

Claims of the Territorial Sea

"Never have national claims in adjacent seas" wrote S. Whittmore Boggs, "been so numerous, so varied, or so inconsistant". The 3 miles limit, strenuously asserted by the United States, the U.K. and others were never universally adopted. The departure had been very few.

Width of the Territorial Sea

Miles	Countries
4 miles	Norway, Sweden, Finland
5 miles	Cambodia
6 miles	Spain, Italy, Yugoslavia, Greece, Uruguay, Lebanon, Cambodia, Iran, Riodeore, Haiti, Ceylon, Israel, Portugal, Thailand.
6.48 miles	Honduras
9 miles	Mexico
10 miles	Albania
12 miles	U.S.S.R., Guatemala, China, Bulgaria, Ethiopia, Indonesia, Libya, Rumania, Saudi Arabia, United Arab Republic, Venezuela (Pakistan and India)
50 k. meters	Chile
200 M miles	Chile, Peru, Ecuador, France, Sansalvador

(b) *Refining Capacity.*

State	Name of Refinery	Installed Capacity (Million Tonnes per Annum)
Bihar	Barauni	3.3
Maharashtra	Bombay (Trombay)	8.8
Assam	Bangagaon	1.0
Kerala	Cochin	3.3
Assam	Digboi	0.5
Assam	Gauhati (Nuamati)	0.8
West Bengal	Haldia	2.5
Gujarat	Koyali	7.3
Tamil Nadu	Madras	2.8
Uttar Pradesh	Mathura	6.0
Andhra Pradesh	Vishakphapatnam	1.5
Capacity due to Addition Processing		8.8
Total		46.6
Projected Capacity by 1990		60 Million Tonnes per Annum

Source: Courtesy Prof. J.N. Nanda, "Development of the Resources of the Sea" p. 72.

APPENDIX C

FISH PRODUCTION

Fish Production (In thousand tonnes)

Year	Marine	Inland	Total
1950-51	534	218	752
1960-61	880	280	1160
1977-78	1515	854	2369
1979-80	1481	855	2336

India has marine fishing fleet	1,50,000.
Traditional Country Craft	60%.
Mechanised Boats	18,000.
Deep Sea fishing vessel in 1980	127.

APPENDIX D

DISTRIBUTION OF MARINE FISHING POPULATION YEAR 1973-77

Total Marine Fish Folk = 1,435,000.
Active Fisherman = 322,532 (22.5 per cent)

State	No. of fishing villages	Total Population	Active Fisherman
1	2	3	4
West Bengal	179	63082	15076
Andhra Pradesh	408	237470	64592
Tamilnadu	374	288586	68317
Pondicherry	21	16414	3785
Kerala	268	391900	80898
Karnataka	145	98832	31740
Goa	40	14865	4067
Maharashtra	299	201423	41539
Gujarat	179	124586	22518
Total	1913	1435158	322532

Source: Planning Commission, National Committee on the Department of Backward Areas, Report on Development of Coastal Areas Affected by Salinity Nov.81.

APPENDIX E

SPECIES-WISE FISH CATCHING IN VARIOUS COASTAL ZONES

Species	Catching in Thousand Tonnes	Zone of Major Fisheries
Plasmo banch	40	West Bengal, Orissa, Andhra Pradesh, Pondicherry, Tamil Nadu, Kerala, Karnataka.
Eal	10	Goa, Maharashtra, Gujarat.
Catfish	50	Goa, Maharashtra, Gujarat, West Bengal, Orissa, Andhra Pradesh, Pondicherry, Kerala, Tamil Nadu, Karnataka.
Chirecentrus	14	Goa, Maharashtra, Gujarat, Pondicherry, Tamil Nadu.
Oil Sardine	218	Pondicherry, Tamil Nadu, Kerala, Karnataka.
Other Sardines	36	West Bengal, Andhra Pradesh, Orissa, Kerala, Karnataka.
Anchoviella	42	West Bengal, Orissa, Andhra Pradesh, Pondicherry, Tamil Nadu, Goa, Maharashtra, Gujarat.
Thrissocles	10	West Bengal, Orissa, Andhra Pradesh, Goa, Maharashtra, Gujarat.
Other Clupeoids	13	West Bengal, Orissa, Andhra Pradesh, Pondicherry, Goa, Maharashtra, Gujarat.
Harpadom Nehereus	122	-do-
Saurida-I-Saurus	10	Pondicherry, Tamil Nadu, Kerala, Karnataka, Goa, Maharashtra, Gujarat.
Hemir Hamphus &	11	Pondicherry, Tamil Nadu.
Belone Elyign Fish	2	-do-
Perches	33	West Bengal, Orissa, Andhra Pradesh, Pondicherry, Tamil Nadu.
Red Mullets	06	Pondicherry, Tamil Nadu, Goa, Maharashtra, Gujarat.
Polynemids	23	Pondicherry, Tamil Nadu.
Sciaemids	128	West Bengal, Orissa, Andhra Pradesh, Pondicherry, Tamil Nadu, Kerala, Karnataka.

APPENDIX E Contd.

Species	Catching in Thousand Tonnes	Zone of Major Fisheries
Ribbon Fish	77	Pondicherry, Tamil Nadu, Kerala, Karnataka, Goa, Maharashtra, Gujarat.
Caranx	23	West Bengal, Orissa, Andhra Pradesh, Kerala, Pondicherry, Tamil Nadu, Karnataka.
Cherinamus	02	-do-
Trachynohy	1.3	Not Ascertained.
Other Carangids	0.1	-do-
Coryphamena	0.6	-do-
Elecatta	0.6	-do-
Leigonathus	31	Pondicherry, Tamil Nadu, Kerala, Goa, Maharashtra, Gujarat.
Cazza	0.1	Not Known.
Lactarius	5.3	West Bengal, Orissa, Andhra Pradesh, Pondicherry, Tamil Nadu, Goa, Gujarat, Maharashtra.
Ponfret	33	West Bengal, Orissa, Andhra Pradesh, Pondicherry, Goa, Maharashtra, Gujarat.
Macheral	86	West Bengal, Orissa, Andhra Pradesh, Pondicherry, Tamil Nadu, Kerala, Karnataka.
Seerfish	26	-do-
Tunnies	23	Pondicherry, Tamil Nadu, Kerala, Karnataka.
Sphaerana	1.8	-do-
Mugil	2.9	West Bengal, Orissa, Andhra Pradesh, Pondicherry, Tamil Nadu.
Bragmaceros	11	Goa, Maharashtra, Gujarat.
Soles	11	West Bengal, Orissa, Andhra Pradesh, Pondicherry, Tamil Nadu, Kerala, Goa, Karnataka, Maharashtra, Gujarat.
Penaed Parawns	86	-do-
Non-Penaed Prawns	97	Pondicherry, Tamil Nadu, Kerala, Karnataka.
Other Crustaceous	70	West Bengal, Orissa, Andhra Pradesh, Pondicherry, Tamil Nadu, Kerala, Goa, Karnataka, Maharashtra, Gujarat.
Miscellaneous	160	-do-

APPENDIX F

MARINE PLACER MINERAL DEPOSITS LOCATED IN THE SEA BED OF THE CONTINENTAL MARGIN

Minerals	Use	Possible Minerable Marine Area
Marine Placers		
Gold	Jewellery, electronics	Alaska, Oregan, California, Philippines, Australia
Platinum	Jewellery, Industry, Iron Ore	Alaska, Black Sea, Russia, Japan, Philippines.
Ilmenite	Source of Titenite	Baltic, Russia, Australia.
Zircon	Source of Zirconium	Black Sea, Baltic, Australia.
Rutile	Source of Titanite	Australia, Russia.
Titanite	Source of Titanite	Australia, Philippines.
Cassiterite	Source of tin	Malaysia, Thailand, Indonesia, Australia, England, Russia.
Monazite	Source of rare earth elements	Australia and United States.
Chromate	Source of chromium	Australia.
Sand and Gravel	Construction	Most continental shelves.
Calcium Carbonate (Aragonite)	Construction, cement agriculture	Bahamas, Iceland, South Eastern United States.
Barium Sulphate	Drilling mud, glass and paint	Alaska.
Diamonds	Jewellery, Industry	South West Africa.
Phosphorite	Fertiliser	U.S., Japan, Australia, Spain, South America, South Africa, India, Mexico.
Potash	Fertiliser Industry	England, Alaska.

Note: Many areas yet to be explored. Hence over all potential is likely to be much more.

APPENDIX G

EFFORTS OF ONGC IN EXPLOITATION OF CRUDE OIL AND INDIA'S REFINING CAPACITY

India and Sri Lanka should have a joint venture in certain locations of Sri Lankan's Coast. Very little petroleum is found in sediments younger than two million years. Ancient continental margines and river deltas are important. Some crude oil reserves are lying under ancient rocks of deltas and buried swamps. India's hopes have increased towards coastal region due to existence of Organic rich shales, parous reservoirs, structural upwarming near continental margines off shore and highseas. It is expected that India will get some oil hydro carbon reserves in near future. Oil and natural gas commission Dehradun, acquired research vessel, Anveshak, which is equipped with seismic, magnetic, geological and geophysical equipments. due to the excellerated efforts of Oil and Natural Gas Commission the refining capacity of India will be 60 million tonnes by 1990. It will surpass national requirement of every year. The year wise requirement and production of petroleum, oil and lubricant are described as below:–

(a) *Crude Oil.*

Year	Demand (Million Tonnes)	Production (Million Tonnes)
1972-73	9.0	5.5
1975-76	10.0	8.0
1978-79	14.0	11.5
1981-82	21.0	16.5
1983-84	30.0	26.2
1990 (Projection)	40.0	Self sufficiency

(b) *Refining Capacity.*

State	Name of Refinery	Installed Capacity (Million Tonnes per Annum)
Bihar	Barauni	3.3
Maharashtra	Bombay (Trombay)	8.8
Assam	Bangagaon	1.0
Kerala	Cochin	3.3
Assam	Digboi	0.5
Assam	Gauhati (Nuamati)	0.8
West Bengal	Haldia	2.5
Gujarat	Koyali	7.3
Tamil Nadu	Madras	2.8
Uttar Pradesh	Mathura	6.0
Andhra Pradesh	Vishakphapatnam	1.5
Capacity due to Addition Processing		8.8
Total		46.6
Projected Capacity by 1990		60 Million Tonnes per Annum

Source: Courtesy Prof. J.N. Nanda, "Development of the Resources of the Sea" p. 72.

APPENDIX H

LATEST PROSPECTS OF OIL STRUCK IN CAUVERY BASIN

September 3, 1989 the Oil and Natural Gas Commission has struck oil once again in the Cauvery Basin — this time at Adiyakkamangalam, five km from Trivarur town in Tamil Nadu's Thanjavur district. On August 31, 1989 oil flowed at the rate of 600 barrels and gas at one lakh cubic metres a day from a well drilled to a depth of 3200 meters an ONGC press release said today.

This wees the fifth discovery of oil and gas since 1985 and was due to ONGS 'Aggressive Exploration Strategy" in the basin.

Following major hydrocarbon strikes at Kovilkalappal, Narimanam and Nannilam in Thanjavur district and Bhuvangiri in South Arcat and a significant offshore discover in the Parto Novo Area. The basis had been upgraded as category, one par with the commercially producing basins like Cambay upper Assam and Bombay offshore.

Indian Express Sept. 04, 1989

APPENDIX I

MINIMUM, MAXIMUM AND AVERAGE PERCENTAGES OF THE MAJOR CONSTITUENTS OF OCEAN FLOOR MANGANESE NODULES

Material	Weight percentages		
	Maximum	Minimum	Average
MnO_2	63.2	11.4	31.7
Fe_2O_3	42.0	6.5	24.3
SiO_2	29.1	6.0	19.2
Al_2O_3	14.2	0.6	3.8
$CaCO_3$	7.0	2.2	4.1
$CaSO_4$	1.3	0.3	0.8
$Ca_3(PO_4)_2$	1.4	traces	0.3
$MgCO_3$	5.1	0.1	2.7
H_2O	24.8	8.7	13.0
Insoluble in HCl	38.9	16.1	26.8

APPENDIX J

MAXIMUM, MINIMUM AND AVERAGE WEIGHT PERCENTAGES OF MAJOR 18 ELEMENTS IN MANGANESE NODULES FROM THE PACIFIC OCEAN

Element	Weight percentages, Statistics on 54 samples		
	Maximum	Minimum	Average
Na	4.7	1.5	2.6
Mg	2.4	1.0	1.7
Al	6.9	0.8	2.9
Sj	20.1	1.3	9.4
K	3.1	0.3	0.8
Ca	4.4	0.8	1.9
Tj	1.7	0.11	0.67
V	0.11	0.02	0.05
Mn	41.10	8.2	24.2
Fe	26.6	2.4	14.0
Co	2.3	0.014	0.35
Nj	2.0	0.16	0.99
Cu	1.6	0.028	0.53
Sr	0.16	0.024	0.081
Zr	0.12	0.009	0.063
Mo	0.15	0.01	0.052
Ba	0.64	0.08	0.18
Pb	0.36	0.02	0.09

BIBLIOGRAPHY

Reference Books

Anti-Kouchine, William, A. and Sternberg Richard, W., "The World Ocean", 2nd Edition, New Jersey, Prentice-Hall Inc., Englewood Cliffs, 1981.

Anon, "Marine Manganese Nodules," Classby, G.P. (Editor), Amsterdom, Elsevier, 1977.

Baily H.S., "The History of Oceanography 1977 – The Voyage of the Challenger in Ocean Series", San Francisco, Freeman.

Bardach John, "Harvest of the Sea", George Allen and Unwin Ltd., London, 1968.

Bhatt, J.J., "Oceanography", New York, D. Van Nostrand Co., 1978.

Bhupinder Singh, Lt. Col. (Retd. Dr.) "Indian Ocean and Regional Security", B.C. Publishers, 47, Ajit Nagar, Opposite Fatima Convent Patiala-147001, Punjab, 1st Edition, 1983.

Carlson and Philbrick, "Geography and World Politics".

Dabrin, Milton, B., "Introduction to Geophysical Prospecting", 2nd Edition, New York.

Doe Brown Book, "Development of Oil and Gas Resources in the United Kingdom", 1977.

Earle A. Sylvia and A.I. Gidding, "Exploring the Deep Frontier", Washington, D.C., National Geographic Society, 1980.

Earney, C.F. Filmore, "Petroleum and Hard Minerals from the Sea", Edword Arnold Ltd, London, 1980.

Edited by R. Gordon Pirie, "Oceanography Contemporary Reading in ocean Science", 2nd Edition, New York, Oxford University Press 1977.

Edward, J. Tarbuck and Fredrick, K., Lutgens, "Earth Science", 3rd Edition, Syndney – Charles E., Merril Publishing Co. 1981.

Encyclopaedia Britannica Vol 2, Published by A Society of Gentlemen in Scotland Inc Welliam Bentan Publisher, London.

Fabian Tract 437 – "Sea Use Planning".

Graham Michael, "Sea Fisheries" 1956, Edevin Arnald, London.

Grabbe David, "The World Energy Book", Edn 1979. MIT Press U.S.A.

Hamlyn, "Encyclopedia of the World in Colour", London p. 31.

Henry S. Parkar, "Exploring the Ocean", Prentice-Hall Inc., Englewood Cliffs, New Jersey.

H.U. Sverdrup, Martin W., Johnson and Richard H. Fleming, "The Oceans Their Physics, Chemistry and General Biology".

Hollister G.D. et. al. Mital Report of "The Deep Sea Drilling Project", Vol II, U.S. Government Printing Office, Washington, 1972.

Horn, Dr. et. al., "World wide Distribution of Manganese Nodules, Ocean Industry", 7 (I), 1972.

J.N. Nanda, "Development of the Resources of the Sea India". Concept Publishing Company, New Delhi-110059, First Published 1989.

Jones, F. and R. Harden (Ed) Sea Fisheries Research 1974, Elek Science, London.

Jones, S. and M. Kumera, Fishes of Laccadive Archipelago, Triveni.

Kent, Peter, E., "Minerals from the Marine Environment", Edward Arnold Ltd, London, 1980.

Maleish W.H. "Oceanography from space", Oceanus 24, 3rd Edition 1981. A Collection of Articles.

Maury J. Hajort, J. & Hell and Hansen B., "The Depth of Ocean Strechert", London, 1912.

Mury M.F., "Physical Geography of the Sea", Haverd New York, 1953, London, 1860, Reprinted Cambridge Mass., 1963.

Mero John L., "Mineral Resources of the Sea", Amsterdam, Elsevier, 1965.

Merit Students Encyclopedia.

"Mineral Resources of the Sea", United Nations Publication E/70/11, New York, 1970.

Murray J. and Renald A.F. "Reports on Deep Sea Deposits in Report on the Scientific Results of Voyage of H.M.S. Challenger During the year 1873-1976, Vol. 5, Eyre and Spettie Words London 1981.

Nelson, S.B., 1971, "Oceanographic Ship for and After", Washington D.C., U.S. Government Printing Office.

P. Green, "The Water of the Sea", D. Van Nastr and Company Ltd., London.

P. Lake, "Physical Geography".

Pearcy and Field, "World Political Geography".

Picard G.L., "Descriptive Oceanography", Pergamon, London 1964.

Qasim, S.Z., Multiple Uses of the Sea, *Everyman's Science 10.*

Qureshi, M.L., "Survey, Economy, Resources and Prospects of South Asia", 1980 Merga Institute Colombo.

Roonwal G.S. "Preliminary Review of Development of Resources of Sea" 1981 – ICWA, New Delhi.

Ross D.A., "Opportunities and Use of the Ocean" Springer Verlag, New York, 1980.

Sandven, Pers, "The Indo-Norwegian Project in Kerala" 1959. Norwegian Foundation for Development Assistance. Oslo.

Schyee S., "The edge of Unfamiliar World – A History of Oceanography", 1973, New York, Dutton.

Setting, M.G. and A. Padmanabhan, "Future of Ocean Resources 1981, – The next 25 years J.S. 1-R. 40, P-708.

Sears M. (Editor) Progress in Oceanography (ann Vols) Oxford, 1963.

Stowek, "Ocean Science, in Edition 1983, New York, Willy.

Stoughton and Hudder, "The Deepest Days", London St. Pauls House Warwick Lane, 1966.

Srimad Bhagwat Geeta.

Sainik Samachar. Vol. 36 No. 6 & 7 Feb 1989 Fleet Review Special. D.P.R.M. of Defence, New Delhi 8.

The Readers Digest Library of Modern Knowledge No. 1, The Readers Digest Association London.

The Readers Digest Great World Atlas. The Readers Digest Association London.

Thurman, Harold V., "Introduction to Oceanography", 3rd Edition, Sundney, Charles E., Merril Publishing Co., 1981.

"Uses of the Sea", Report of the Secretary General, United Nations Economic and Social Council. Publication E/1980/68.

Valkenburg Samuel Van and Stotz Carl, "Elements of Political Geography", pp. 113-114, Prentice-Hall Inc. Enalwood Cliffa N.J. 2nd Edition.

Ved Vyas, *Mahabharat Geeta.*

Vetter, Richard C. (Ed) 1974, "Oceanography", the last Frontier voice of America – Farum series.

Editor William P., Hlsey, "Merit Students Encyclopedia" Vol 14, Macmillan Educational Corporation, New York.

Weyl, P.K. "Oceanography an Introduction to Marine Environment", 1970 – New York, Willy.

Periodicals, Pamphlets, Reports and Dissertations

Anm, 1987, "India Opens Door to Joint Ventures", *Fishing News International*, June 1982.

Annual Report 1987-88, Department of Ocean Development, Government of India, New Delhi.

Annual Report 1988-89, Department of Ocean Development, Government of India, New Delhi.

Article of Dr. M. Ravindran on "Deep Sea Mining", *Ocean* Engineering Centre, I.I.T. Madras.

Article of Dr. S.P. Subramanian, "Geological Oceanography", Ocean Engineering Centre, I.I.T. Madras.

Article of Prof. V.S. Raju, "Introductory Lecture on Ocean Resources and Ocean Engineering Aspect". Ocean Engineering Centre, I.I.T. Madras.

Article of Dr. A.K. Malhotra, "Offshore Activities Related to Oil and Gas in India". International consultant Formerly Member of O.N.G.C., Bombay.

Article S. Qasim, "A Technological Forecast of Ocean Research and Development in India", Courtesy : Impact of Science on Society, 1983, No. 314 (P.P. 38-52), Published in *Strategic Digest*, March 1985, Vol XV, No.8, Institute for Defence Studies and Analyses, New Delhi.

Bennet, R. and G.C. Eddie, 1968, 'The Role of Commercial Vessels in Fisheries Research and Development,' 2nd FAO Tech Conference on Fisheries Research craft, seattle No. 7(iii) FAO Rome pp. 9.

Editorial in the *Tribune*, June, 22, 1982.

Exploration of Manganese Nodules in the Indian Ocean, Suggested approach for R & D, August 1981. National Institute Oceanography.

Exploration and Exploitation of Marine Minerals, Oct. 1981 (No 10).

Calvane Eric (Eaulard Rov) Sea Technology. (Nov 1983) P.P. 35-36.

Dwivedi, S.N. and K.B. Sriramamurthy, 1979. Oil Pollution and its consequences with special reference to Bombay Coast. Contributions to Marine Science dedicated to Dr. C.V. Kurian, Cochin University.

Frazer, J.N. and L.L. Wilson 1980, Management Nodules Researches in the Indian Ocean, *Marine, Mining* 2(34) 257-292.

Gullond, J.A. and J.E. Cerros, 1968, "Management of Fisheries Resources, *Adv. Mor. Biol* 6-171.

Pamphlet, "Polymetallic Nodules India's Work on the Resources for the Future," Department of Ocean Development, Government of India, New Delhi.

Pamphlet, "Activities of the Department of Ocean Development in the Field of Ocean Science and Technology", Department of Ocean Development, Government of India, New Delhi.

Pamphlet, "India and the Antarctica Scenario", Department of Ocean Development, Government of India, New Delhi.

Qasim, S.Z. 1980, Role of Aquaculture in Rural Development, Invention Intelligence : Special Number 12 p. 107-112.

Rechmitzer Andreas, "Rov Technology A Ticket to Scientific Discovery," *Sea Technology* (Dec, 1, 1985) pp. 65.

Science page, "Recovery of Deep Treasure", *Times Magazine* (Nov. 1987), p. 54.

Siddiquie, H.N., D.A. Das Gupta, N.R. Sen Gupta, P.C. Srivastava and T.K. Malik 1978. Manganese Iron Nodules from the Indian Ocean, *Ind. J. Mor. Sc.* 7 : 239-253.

Verghese, G. 1974. Shark Resources of the Laccodive Waters. *Seafold. Exp*, J. 6-65-68.

Wave Energy Oscillator : A New Concept in shore Protection and Coastal Engineering. Tech. Report 05/80 Sept. 1980.

Wave Power : A Status Report, Sept 1981.

World Fisheries and the Law of the Sea, 1981, pp. 55. U.N. Publications.

INDEX